Studies in Scientific Realism

Studies in

SCIENTIFIC REALISM

André Kukla

New York   Oxford
OXFORD UNIVERSITY PRESS
1998

Oxford University Press

Oxford   New York

Athens   Auckland   Bangkok   Bogota   Bombay   Buenos Aires
Calcutta   Cape Town   Dar es Salaam   Delhi   Florence
Hong Kong   Istanbul   Karachi   Kuala Lumpur   Madras
Madrid   Melbourne   Mexico City   Nairobi   Paris   Singapore
Taipei   Tokyo   Toronto   Warsaw

and associated companies in
Berlin   Ibadan

Published by Oxford University Press, Inc.
198 Madison Avenue, New York, New York 10016

Oxford is a registered trademark of Oxford University Press

Library of Congress Cataloging-in-Publication Data
Kukla, André, 1942–
Studies in scientific realism / André Kukla.
p.   cm.
Includes bibliographical references and index.
ISBN 0-19-511865-0
1. Realism.   2. Science—Philosophy.   3. Science—Methodology.
I. Title.
Q175.32.R42K85   1998
501—dc21   97-27726

9 8 7 6 5 4 3 2 1
Printed in the United States of America
on acid-free paper

To Kaila, for the gift of balance

Preface

The title of this book is somewhat misleading. *Studies in Scientific Realism* suggests a series of discontinuous investigations, wherein the author turns from one issue to another as different aspects of the topic strike his fancy. In fact, I've tried to provide a systematic and exhaustive survey of all the important considerations that bear on the question whether the unobservable entities postulated by science can be known to exist. The most precisely correct title would have been *Realism and Antirealism in the Philosophy of Science*. Unfortunately, a book with that title appeared just as I was putting the finishing touches on my own manuscript. Other, terser candidates, such *Scientific Realism*, *Science and Reality*, and *Reality and Science* are also taken. So there it is.

A substantial portion of this material has been culled from prior journal articles. However, it's been rewritten, supplemented with new insights, and reorganized for continuity to the point where it's no longer possible to line up specific chapters with specific articles. Here's a list of the articles that were mined:

"On the coherence of instrumentalism," *Philosophy of Science*, 1992, 59, 492–497. Reprinted by permission of the University of Chicago Press. Copyright 1992 by the Philosophy of Science Association. All rights reserved.

"Non-empirical theoretical virtues and the argument from underdetermination," *Erkenntnis*, 41, 1994, 157–170. Reprinted with kind permission from Kluwer Academic Publishers.

"Scientific realism, scientific practice, and the natural ontological attitude," *British Journal for the Philosophy of Science*, 45, 1994, 955–975. Reprinted by permission of Oxford University Press.

"Scientific realism and theoretical unification," *Analysis*, 55, 1995, 230–238. Reprinted by permission of the author.

"The two antirealisms of Bas van Fraassen," *Studies in History and Philosophy of Science*, 26, 1995, 431–454. Copyright 1995, reprinted with kind permission from Elsevier Science Ltd., The Boulevard, Langford Lane, Kidlington OX5 1GB, UK.

"Antirealist explanations of the success of science," *Philosophy of Science, 63* (Supplement), 1996, S298–S305. Reprinted by permission of the University of Chicago Press. Copyright 1996 by the Philosophy of Science Association. All rights reserved.

"Does every theory have empirically equivalent rivals?", *Erkenntnis, 44,* 1996, 137–166. Reprinted with kind permission from Kluwer Academic Publishers.

"The theory-observation distinction", *Philosophical Review, 105,* 1996, 173–230. Reprinted by permission of the publisher.

I thank the following people whose e-mail, snail-mail, and face-mail have helped to clarify my ideas about scientific realism: Ron Amundson, James Brown, Danilo Chiappe, Christopher Green, Rebecca Kukla, Larry Laudan, Stephen Leeds, Jarrett Leplin, Margaret Morrison, and Bas van Fraassen. The Social Sciences and Humanities Research Council of Canada has provided research support during the several years it took me to complete this project.

Toronto, Ontario   A.K.
July 1997

Contents

Studies in Scientific Realism

The Varieties of Realism

The first order of business is to distinguish among several varieties of realism and antirealism. This unexciting but essential preliminary task is made all the more tedious by the lack of a standardized nomenclature in the field. Take "instrumentalism". In the current philosophical literature, this term is often used to refer to a form of antirealism which denies that theoretical statements have truth-values. Almost as often, however, it covers any form of antirealism about theoretical entities. In the first sense of the term, van Fraassen isn't an instrumentalist; in the second sense, he is. In this chapter, I present a taxonomy of realisms and antirealisms that will (usually) be adhered to throughout this book. Current usage is taken into account as much as possible, but I've had to make some arbitrary decisions. I've even had to introduce a couple of neologisms. The resulting scheme is adequate for the purpose of locating the present work in its broader philosophical context. Perhaps it will also encourage terminological standardization among other workers in the field. It shouldn't be expected that my scheme will be exhaustive of all the forms of realism and antirealism that have occupied the attention of philosophers. Those that I deal with do, however, constitute a set of core positions. They're the "main sequence" of realist and antirealist views in our era.

1.1 The Vertical Varieties of Realism

Realisms and antirealisms can be divided *vertically*, by the tenuousness of the objects to which they allude, or *horizontally*, by the thesis asserted or denied of these objects. Some realisms and antirealisms also have philosophically interesting differences of *degree*. In this section, the focus is on the vertical distinctions. Consider the following sequence of existential hypotheses:

1. Sense-data, like "being appeared to greenishly", exist.
2. The commonsense objects of perception—sticks and stones—exist.

3. The unobservable entities postulated by scientific theories, such as electrons and unconscious mental processes, exist.
4. Timeless abstract entities such as numbers, sets, and propositions exist.

The several vertical realisms and antirealisms to be defined correspond to different conjunctions of these hypotheses and their negations. Actually, the hypothesis that a certain type of entity *exists* is only one of the several types of horizontal
realisms to be differentiated in section 1.2. Until then, the import of these existential
claims should not be pondered too closely. They should be treated rather as place-
holders for any of the horizontal distinctions to come.

There's a tendency for espousers of any one of these vertical theses to accept all
the others that are lower in the hierarchy. Believers in the reality of numbers don't
usually balk at accepting electrons, and scientific realists who believe in electrons
don't usually entertain misgivings about sticks and stones. Nevertheless, each of the
four hypotheses is logically independent of the other three. Indeed, there are more
than a few historical examples of people whose views don't fit into the hierarchical
mold. The most glaring case is the not at all uncommon position that accepts the
reality of sticks and stones, but denies that there are such things as sense-data—that
is, position (–1 & 2). A more extreme departure from the hierarchical scheme is represented by Plato, according to whom *only* abstract objects are real (–1 & –2 &
–3 & 4).

The current name for proposition 4 is *Platonism*. Platonism should not be con-
fused with the philosophy of Plato. Unlike Plato, contemporary Platonists don't assert that *only* abstract objects exist—all the contemporary Platonists that I know of
are comfortable with both stones and electrons. When I want to refer to Plato's view,
(–1 & –2 & –3 & 4), I'll call it *pure* Platonism. Proposition 3 is *scientific realism*, and
proposition 2 is *commonsense realism*. *Phenomenalism* is not simply proposition 1—
it's the thesis that *only* sense-data exist—that is, (1 & –2 & –3 & –4).[1] It's useful to
have a name for proposition 1 by itself, without the phenomenalist implication that
sense-data are more real than sticks and stones. I suggest *sense-data realism*. Sense-
data realism is to Platonism as phenomenalism is to pure Platonism.

This book is about the debate that has dominated realist-antirealist discussion
in the past generation. This debate is conventionally described as a battle between
scientific realists and antirealists, but that isn't entirely accurate—at least not according
to the definition of scientific realism given above. If "scientific realism" is a name for
proposition 2, and if, as etymology suggests, antirealism is the contradictory of realism, then pure Platonists and phenomenalists would have to be counted among the
scientific antirealists. In fact, the status of neither sense-data nor abstract entities is
at issue in the current debate about scientific realism. Moreover, the existence of
commonsense objects isn't in question either—it's a presupposition of the discussion that sticks and stones exist. If we're to continue to call this debate a battle between scientific realists and antirealists, we have to say that "scientific realism" is the
name not just for proposition 3, but for the conjunction of 2 and 3—and that scientific antirealism isn't just the negation of 3—it's the conjunction of 2 and the negation of 3. For the issues I want to discuss, I find it convenient to adopt these definitions most of the time. Sometimes, however, I will want to refer to proposition 3 by

itself. I'll use "scientific realism" for this purpose as well. The resulting ambiguity is enshrined in current usage and has proven itself to be innocuous. In any case, in the more elaborate of their two senses, scientific realism and antirealism aren't contradictories at all. They're contraries, etymology notwithstanding.

Since the two contending parties are contraries rather than true contradictories, there is a difference between arguments *for scientific realism* and arguments *against scientific antirealism.* The former, if sound, establish the falsehood of antirealism, but the latter don't establish the truth of realism. For example, the argument that there's no coherent way to distinguish theoretical entities from the commonsense objects of perception may prove to be the undoing of scientific antirealism.[2] But even if this argument is sound, it doesn't establish that scientific realism is true, for a proof that commonsense objects and theoretical objects are in the same boat is also compatible with phenomenalism or pure Platonism. By the same token, there's a difference between arguments for antirealism and arguments against realism. The former, if sound, establish the falsehood of realism, but the latter don't establish the truth of antirealism. However, these distinctions collapse when scientific realists and antirealists argue with each other out of earshot of any phenomenalists or pure Platonists. Their shared presupposition that commonsense objects exist turns a disproof of antirealism into a proof of realism, and a disproof of realism into a proof of antirealism.[3] For instance, if it's true that you can't distinguish commonsense objects from theoretical objects, *and* if commonsense objects exist, then of course so do theoretical objects.

Since this book deals exclusively with the debate between scientific realists and antirealists, I will also adopt the presupposition that commonsense objects exist. I wish it to be understood, however, that this assumption is purely tactical. In fact, I think that phenomenalism, which entails the negation of both scientific realism and antirealism, is a candidate to be reckoned with. Here's a prima facie case for that claim. Most of the arguments traded between realists and antirealists of any stripe involve a comparison between two adjacent levels in the fourfold chain of beings: realists about level n want to pull antirealists up from level $n - 1$, and antirealists about n want to drag the realists down from n to $n - 1$. There are three issues of this type, corresponding to $n = 2$, 3, and 4. Not surprisingly, the arguments that crop up in these three debates have many formal similarities with one another. The antirealist who wants to stay at the lower level typically appeals to parsimony and epistemic safety. Conversely, realists on the higher level complain either that the lower level lacks the conceptual or explanatory resources needed for doing a necessary job, or that in purporting to do the job, the antirealist inappropriately co-opts the resources of the higher level without acknowledging it. In making these claims, the disputants appeal to general philosophical principles that justify either antirealist restraint or realist expansiveness. If these principles are general enough, there's a good chance that they will apply equally well to movement up from *any* level $n - 1$ to level n, or down from any n to $n - 1$. Here's an example of what I mean.

Several scientific realists have argued for moving from level 2 to level 3 on the ground that scientific antirealists tacitly employ the conceptual resources of level 3 in formulating their beliefs. Let T^* be the thesis that the empirical consequences of theory T are true. T^* in effect says that the observable world behaves *as if* T were true, but it doesn't assert that T is true. Van Fraassen and other scientific antirealists have maintained that there can be no rational warrant for believing in the theoretical implications of T that go beyond the content of T*. The argument is that T* gives us all the predictive power of T. Moreover, since T* is a logical consequence of T, it must be at least as likely as T. Therefore, since there's nothing to be gained by adopting the surplus content of T, we should regard T* as our total account of the world. However, Leplin and Laudan (1993) note a putative deficiency in T* that, they claim, warrants movement up to T after all. Their charge is that "[T*] is totally *parasitic* on the explanatory and predictive mechanisms of T'" (13, emphasis added).[4] That is to say, you can't even describe what the content of T* is without alluding to the theoretical entities posited by T. You can't claim that the observable world behaves as if electrons exist without mentioning electrons. Clendinnen (1989) makes it clear that this is supposed to be a reason for *believing* T if you believe T*:

> The predictive content of [T*] is parasitic on [T], and this bears on the rationale we could have for accepting or believing the former. If we doubt the truth of [T] we have no basis to expect the predictions made by it. . . . The rationale for our predictions is the confidence we have in our speculation about hidden structures. [T*] might be marginally more probable than [T], but in doubting the latter we abandon all grounds for confidence in the former. (84)

The soundness of this argument will be assessed in due course. For the present I wish only to note the close parallel between this critique of scientific antirealism from the perspective of scientific realism, and the following critique of phenomenalism from the perspective of commonsense realism. Here is Lenn E. Goodman (1992) on the necessity of moving from level 1 to level 2:

> Consider the phenomenalist's paradigm statement: "I am appeared to tea-mintily": . . . There is no cup, no tea in fact, presumed within the judgment, only the expression of a single subjective impression, a model for all other subjective impressions. . . . But what does it mean to identify a distinctive sensation here, in the tea-mintiness of the present sensation? Can the sensation be named without reference to the object language? That is a reference the phenomenalist . . . freely makes: Uninhibitedly, he refers to tea, an object in the world. . . . The phenomenalist uses particular transcendental terms and universal lawlike predicates to demarcate and designate what he means. . . . [The phenomenalist's] usage is wholly *parasitic* upon that language. That is where it gets its specificity and the distinctness of its notions. That is where it goes to vindicate and differentiate its claims. . . . So where is the economy? (245–246, emphasis added)

Evidently, T is to T* in the 2-to-3 debate as commonsense physical description is to phenomenal description in the 1-to-2 debate. I don't know how powerfully this parasitism argument strikes the uncommitted reader on first reading. In chapter 5, I argue that it's not at all compelling. But both sides of both realism debates must surely agree that the arguments are equally compelling in both debates. If you think that the parasitism argument is reason enough to move from level 1 to level 2, then you pretty much have to keep moving, for the same reason, from level 2 to level 3. And if you think that the parasitism argument is insufficient reason for a commonsense realist to become a scientific realist, then you can't very well take the parasitism of their descriptions to be a telling reproach against the phenomenalists.

Moreover, aren't the foregoing arguments the 1-to-2 and 2-to-3 counterparts of Putnam's (1975a) indispensability argument for going from 3 to 4? According to Putnam, the fact that scientific theories require mathematics for their formulation is reason enough to believe in the existence of mathematical objects. It's difficult to think of a good reason for accepting any one of these three arguments and rejecting either of the other two (assuming, of course, that the claims about parasitism and indispensability are true). So it seems that if you eschew phenomenalism because of the parasitism of phenomenalist descriptions, you may be committing yourself both to scientific realism and to Platonism.

Here's another example of the same sort of thing. Benacerraf (1965) has argued against Platonism about numbers on the grounds that there are indefinitely many conflicting ways to identify numbers with sets. The von Neumann ordinals and the Zermelo ordinals (inter alia) are equally viable set-theoretical conceptions of numbers. Thus, Benacerraf maintains, we have no principled reason to say that numbers really are one or the other. Benacerraf's reason for refusing to move from level 3 to level 4 has an obvious counterpart in the 2-to-3 debate: the underdetermination of scientific theories by all possible empirical data. According to this argument, the fact that every theory has empirically equivalent rivals entails that we have no basis for believing any one of them. The merits and demerits of the underdetermination argument are examined in chapters 5 and 6. Whatever our final assessment of the argument may be, however, it seems compelling that we have to pass the same judgment on Benacerraf's argument (assuming that both scientific theories and numbers really are underdetermined by their evidential bases). If you think that anti-Platonism about numbers follows from the fact that there are indefinitely many set-theoretical formulations that capture all numerical properties, then you should also refuse to believe in the theoretical entities of science because of the fact that there are indefinitely many empirically equivalent rivals to any scientific theory.

This line of thinking suggests a possible avenue for effecting at least a partial resolution of the scientific realism-antirealism debate, even if it proves impossible to come up with a direct demonstration of either hypothesis that's persuasive to the other side. It may be that all the important arguments, inconclusive as they may be, for moving from level $n - 1$ to level n, or for eschewing such a move, apply with equal force to all n. Such a state of affairs would still not tell us precisely which entities exist. But it would tell us that the two midpoints in the hierarchy—belief in commonsense objects but not in theoretical entities, and belief in theoretical entities but not in abstract entities—are untenable stopping points. That is to say, it's possible that any impetus we might have for going from sense-data to sticks and stones would propel us all the way to numbers, and conversely, any reason for refusing to move from electrons to numbers might precipitate us backward all the way to phenomenalism. The upshot would be that the only tenable positions are phenomenalism and an omnibus ontology that contains all four sorts of entities. Either way, propositions 2 and 3 would have to have the same truth-value. Thus, scientific antirealism, defined as adherence to (2 & –3), would stand refuted, while scientific realism, defined as adherence to (2 & 3), would still be in the running as a fragment of the omnibus ontology. To be sure, this is only an idea for an argument. But my guess is that it can be made to work. Speaking autobiographically, my current state of opinion on the scientific realism issue can be represented as (2 & 3)V(–2 & –3). I think that commonsense objects and theoretical objects are in the same boat, but I'm not yet sure what that boat is.

I now set this argumentative strategy aside and return to the debate between the two midpoints in the hierarchy of beings. I've raised the broader topic only to emphasize that even if one side were to win the circumscribed skirmish between scientific realists and antirealists, it wouldn't necessarily have nailed down an absolute victory. Under the terms of engagement in this debate, scientific realism wins if it comes up with a persuasive proof that proposition 2 entails proposition 3. This is, of course, compatible with 2 and 3 both being false. Similarly, scientific antirealism wins if it can be shown that proposition 2 entails the *negation* of proposition 3, which is also compatible with 2 and 3 both being false. Thus, either "victory" can be accommodated by phenomenalism and pure Platonism, as well as by several other points of view that are too exotic to be named.

1.2 The Horizontal Varieties of Realism

In this section, I distinguish between *semantic, metaphysical,* and *epistemic* realist theses. I don't think that my definitions are eccentric, but I lay no claim to having captured any common or influential usage. In particular, my realisms don't line up at all well with Horwich's (1982) "semantic", "metaphysical", and "epistemological" realisms. I'll define the horizontal theses first regarding theoretical entities.

Semantic realism (really semantic scientific realism) is the view that statements about theoretical entities are to be understood literally: "Electrons are flowing from point A to point B" is true if and only if electrons are indeed flowing from point A to point B. To assert this much is not yet to be committed to the view that electrons ever *do* flow from one point to another, or that electrons exist, or that we would ever be able to discover that electrons exist even if they do. Semantic realism is logically the weakest of the three horizontal realisms. The thesis that denies it — *instrumentalism* — is therefore the strongest horizontal antirealism. According to instrumentalists, the question whether electrons exist isn't even a well-formed question. Instrumentalists regard theoretical terms as uninterpreted tools for systematizing observations and making predictions.

Reductionism is the view that theoretical entities are "constructions" out of more familiar materials — canonically, that theoretical terms can be defined in language that refers exclusively to commonsense objects. Whether reductionism is deemed to be a realist view or an antirealist view depends on what it's compared to. On the one hand, it entails what we've called semantic realism: if statements about theoretical entities are equivalent to statements about commonsense objects, then, since the latter are truth-valuable, so must the former be. On the other hand, reductionism violates a realist sentiment — or suspicion — to which we haven't yet given a name. The scientific realist's suspicion is that the world of perceptible objects is smaller than the whole world. Let's call this thesis *metaphysical realism*. According to metaphysical realists, there may very well exist theoretical objects that are not part of the observable world. If electrons are such objects, then reductionism doesn't apply to electrons. To deny metaphysical realism is to claim that the world and the observable world are one. A reasonable name for this position is *positivism*. Instrumentalism and reductionism are both species of positivism—for the world and the observable world are one if theoretical statements are non-truth-valuable, or if they're reducible to observation statements. Some other conceptual connections are that metaphysical realism (as well as reductionism) entails semantic realism, and reductionism may be rejected on either instrumentalist or metaphysical realist grounds.

Epistemic realism is the thesis that we can come to *know* that theoretical entities exist. Actually, this is only one of several grades of epistemic realist claims. Once again, I use a single formulation as a temporary stand-in for any member of a broader class to be delineated later. Epistemic *anti*realism, the view that the existence of theoretical entities *cannot* be known, is also called *constructive empiricism*. Its foremost proponent—and the coiner of the phrase—is Bas van Fraassen (1980). Van Fraassen and others use "constructive empiricism" just as defined some of the time. But the same term is also used to refer to variably larger samples of van Fraassen's philosophy. A commonly included item is the claim that scientists adopt—or need only adopt—a special attitude of "acceptance" toward scientific theories, which is distinct from both belief and utter indifference. So not every refutation of constructive empiricism in the broad sense is a refutation of constructive empiricism in the narrow sense. Unless otherwise indicated, I'll use the term "constructive empiricism" to mean the negation of epistemic realism.

If statements about electrons are not truth-valuable, then one surely can't have any knowledge about electrons. So, constructive empiricism is entailed by instrumentalism. The entailment doesn't run the other way, however, for one can agree that "electrons exist" is truth-valuable and deny that we can ever have adequate grounds for deciding that it is true. In fact, van Fraassen explicitly endorses semantic realism, and often includes it in the package that goes by the name of "constructive empiricism". However, semantic realism plays no essential role in development of van Fraassen's other views about science. It's merely a concession to the powerful considerations in its favor. These considerations will not be rehearsed here.

The horizontal realisms and antirealisms are orthogonal to the vertical options discussed in section 1.1. Thus, there are several varieties of phenomenalism: (1) instrumentalist phenomenalism, which claims that physical-object language is just a tool for organizing and predicting sense-data; (2) reductionist phenomenalism, which asserts that physical-object language can be translated into sense-data language; and (3) constructive-empiricist phenomenalism, according to which we could never know that physical objects exist, even if they do. Positivist phenomenalism is the disjunction of 1 and 2. There are also three kinds of anti-Platonism. Constructive-empiricist anti-Platonism is the view that even if numbers and other abstract entities exist, we could never know it. Arguments for this position have been made on the basis of a causal theory of knowledge: since abstract entities are not in space or time, they can't enter into causal relation with us; hence, by the causal theory of knowledge, we can have no knowledge of them, even if they exist (Benacerraf 1973; Field 1980).

At this point, I bid farewell to semantic and metaphysical realism, to their several antirealisms, and to Platonism and phenomenology. Henceforth, the topic is solely epistemic scientific realism and its contrary, epistemic scientific antirealism (a.k.a. constructive empiricism about theoretical entities). Given this restriction of scope, I feel free to abbreviate these labels to "realism" and "antirealism", respectively.

1.3 The Degrees of Epistemic Realism

Of the three horizontal realisms, epistemic realism is the only one that permits philosophically interesting degrees. What would degrees of metaphysical realism be like? Would a high degree of it mean that there are very many unobservable entities? But nobody cares whether only a few theoretical terms denote transcendent entities, or whether many of them do. The only philosophically interesting question is whether there are any unobservable entities at all.[5] The situation is different with epistemic realism. Here are four grades of epistemic realism worth distinguishing. These distinctions of epistemic degree can be made regarding any of the vertical realisms. However, I spell them out in full only for the case of theoretical entities.

The first and strongest view is that we know that our best current scientific theories are true. Against this view is Putnam's (1978) "disastrous induction": all our past scientific theories have come to be regarded as false; hence it is overwhelmingly likely that future science will come to regard our current theories as false. This is indeed a reason for not endorsing the view that our current theories are true without a second thought. Like all inductions based on bare enumeration, however, the force of its conclusion is underwhelming. I have more to say about the status of the disastrous induction in chapter 2. Here, let's observe only that it doesn't close the door on the possibility that our best current theories are known to be true.

Rightly or wrongly, the disastrous induction has impelled many epistemic realists to weaken their claim to the following: what we know is that our best current theories are *close* to the truth. This formulation of epistemic realism has been exhaustively discussed — and severely criticized — by Laudan (1981). The main problem with it, according to Laudan, is that the concept of approximate truth isn't nearly clear enough to carry the philosophical burden that it's being made to bear. Moreover, if the theoretical terms of approximately true theories have to refer to real entities, then the weaker second thesis is subject to the same disastrous induction as the first — for the ontologies of the best theories of other ages have routinely been overturned by later science (phlogiston, caloric, the ether, absolute space, etc.).

A third version of epistemic realism has it that we're *rationally warranted in believing* that our best current theories are true, or in believing that they're close to the truth. This move from knowledge to rationally warranted belief puts the thesis beyond the reach of any simple disastrous induction, for it's not at all obvious that there can't be a temporal series of theoretical beliefs, the changes among them precipitated by empirical discoveries and conceptual innovations, such that (1) every member of the series is false, yet (2) belief in every one of them was rationally warranted in its time. It can even be argued that this brand of epistemic realism is immune to the problem that Laudan complains most about — the vagueness of the concept of approximate truth—for we're surely warranted in doing the best we can with vague concepts *some* of the time.

An assessment of the foregoing view would require us to ascertain whether belief in our best current theories is in fact warranted. I don't wish to take on this evaluative burden. The epistemic realism that I want to talk about is even weaker than the third variety. It asserts only that *it's logically and nomologically possible to attain to a state that warrants belief in a theory*. Maybe it's crazy to believe in *any* of the theories that scientists have come up with so far. But at least it's possible that science will one day come up with a theoretical proposal that warrants our tentative assent. This is the view that Leplin (1997) has dubbed *minimal epistemic realism*. It's a very weak doctrine, as far as epistemic realisms go (though I show in chapter 5 that, contra Leplin, it isn't the weakest hypothesis that still qualifies as a brand of epistemic realism). It follows that its negation is a very strong form of epistemic antirealism—it's that *we can never have adequate grounds for believing in any theory*. This is the grade of epistemic antirealism that van Fraassen espouses under the name of "constructive empiricism", and it's what I mean by "constructive empiricism" from now on. In fact, I hereby give myself license occasionally to use a simple "realism" or "antirealism" in lieu of "minimal epistemic realism about theoretical entities" and its constructive-empiricist negation.

1.4 Things to Come

The rest of this book is devoted to an examination of the debate between minimal epistemic realists about theoretical entities (henceforth, realists) and constructive empiricists (antirealists). My conclusion will be that neither side has gained a decisive advantage over its adversary. In fact, recent realist-antirealist discussions have displayed a tendency to admit—or at least to conceive of the possibility—that neither side may have the resources to persuade a rational proponent of the other camp to change her ways. The various morals that have been drawn from this state of affairs are discussed in the last chapter. The view that I endorse in the end is very similar to van Fraassen's (1989) most recent position—namely, that the differences between realists and antirealists are indeed irreconcilable, but that espousal of either doctrine is nevertheless irreproachable. Van Fraassen, quite irreproachably, conjoins this view with antirealism; I don't. There's not much of interest to be said about this particular difference of opinion. The more interesting difference is that I don't think that the thesis of irreproachable irreconcilability requires the radical epistemological novelties that van Fraassen advocates. This is the sober note on which I close my investigation.

Realism and the Success of Science

T he bedrock of the case for scientific realism is the argument from the success of science. Most versions of the argument have the following structure:

(SS1) The enterprise of science is (enormously) more successful than can be accounted for by chance.

(SS2) The only (or best) explanation for this success is the truth (or approximate truth) of scientific theories.

(SS3) Therefore, we should be scientific realists.

I'll refer to this argument schema as SS. The canonical version of SS is Putnam's (1975a), according to which the success of science (SS1) would be a "miracle" if scientific theories were not true. To say this is to claim that theoretical truth is the *only* available explanation for the success of science. Putnam's version is therefore a particularly fragile form of the argument: the formulation of a single coherent alternative explanation would be enough to block the conclusion. A more robust argument is that we should be realists because theoretical truth (or approximate truth) is the *best* explanation for the success of science.

By the "success of science", I mean that our scientific theories enable us to make significantly more correct predictions than we could make without them. This formulation is more circumscribed than, but roughly consistent with, Laudan's broadly pragmatic characterization:

> A theory is successful provided that it makes substantially correct predictions, that it leads to efficacious interventions in the natural order, or that it passes a suitable battery of standard tests. (1984b, 109)

There's nothing of central significance in this chapter that hinges on the differences between these two formulations.

Antirealist objections to SS fall into three groups. It's been claimed (1) that neither truth nor approximate truth qualifies as a passable explanation for the success of science; (2) that even if truth or approximate truth did explain the success of science, there are also comparably good — perhaps even better — antirealist explanations; and (3) that even if truth or approximate truth were the only or the best explanation for the success of science, the conclusion that we should be realists wouldn't follow from the premises without additional and question-begging assumptions. I examine each of these claims in turn. But first I want to draw attention to a fourth type of objection that is conspicuously absent from the antirealist literature.

2.1 The Missing Counterargument

Antirealists have never tried to deny the truth of the first premise, SS1. This absence is not to be explained by the unthinkability of denying that science is successful. Indeed, the denial of SS1 is a staple of social constructionist and other relativist analyses of science. According to Latour and Woolgar (1979), for instance, the *apparent* success of science is explained by the fact that scientists *construct* the data that confirm their own theories. This process of self-validation might be said to result in successful theories on some notions of success — perhaps even on Laudan's notion — for a self-validated theory does correctly predict the data that will be constructed to confirm it. But it's clear that neither realists nor antirealists, Laudan included, intend that this kind of performance should count as a scientific success for the purpose of SS. The reason is that a self-validating success doesn't depend on any of the intrinsic properties of the *theory*. Our predictions would be just as successful with any other theory that was equivalently situated in our social milieu. The characterization of success that I proposed above gives the right judgment for this case: self-validated theories aren't successful because they don't enable us to make more correct predictions than we could make with any *other* theories that were appropriately embedded in our social life.[1]

Despite its conceivability, antirealists have never tried to undermine SS by denying SS1. Given the well-known propensity of philosophers to occupy all possible argumentative niches, this is a cause for minor wonderment. The absence of this particular strategy would be understandable if the thesis of antirealism *presupposed* the success of science. But in fact SS1 is not entailed by antirealism, or by realism, or by the disjunction of antirealism and realism. There is no logical impropriety in endorsing constructive empiricism — or epistemic realism — and claiming that all theories that have been suggested to date — or that will ever *be* suggested, for that matter — are dismal empirical failures. What realists and antirealists must presuppose (and what most relativists deny) is that it's *nomologically possible* for us to have successful theories. The issue between realists and antirealists wouldn't make any sense unless empirical success were at least possible. There's no point arguing about whether empirical success warrants theoretical belief if empirical success is impossible. But the success doesn't have to be actual. In principle, the debate about scientific realism could have been conducted by cavepersons long before the first remotely successful scientific theory was conceived.

Since SS1 is independent of antirealism, why haven't any antirealists tried to refute the realist argument *from* the success of science by denying that science is successful? Here is one likely reason: argument SS, which is about *actual* theories, can be reformulated as an argument about *possible* theories in such a way that the new argument is at least as plausible as the old. Antirealists may not have to concede that any theory that has ever been formulated (or that will ever be formulated) is successful. But they do have to admit that it's *possible* for us to come into possession of a successful theory. And then the realist can claim that it's only the truth of such a theory that would explain its success. The new argument for realism—call it SS'—would look like this:

> (SS1') Some nomologically accessible theories would be successful if we used them to make predictions.

> (SS2') The only (or best) explanation for this success would be the truth or approximate truth of the theories in question.

> (SS3') Therefore, there are nomologically possible circumstances—namely, the circumstance of being in possession of a successful theory—wherein we would be justified in believing that a theory is true.

Argument SS' is an *existential generalization* of SS. If SS is sound, then so is SS', but the refutation of SS doesn't automatically sink SS'. That is to say, SS' is the stronger argument. In particular, the negation of SS1 would show SS to be unsound, but it would not, by itself, adversely affect the status of SS'.

Quite likely, Latour and Woolgar and other relativists would wish to deny SS1' as well as SS1—and this would undermine SS' as well as SS. The truth of SS1', however, is not up for grabs in the debate between realists and antirealists. But the fact that SS1 is independent of both realism and antirealism means that realists and antirealists are both free to avail themselves of any argumentative strategies that might involve its negation. I discuss one such strategy in the next section. Ironically, it's deployed in a *realist* defense of the generalized argument for the success of science SS'.

2.2 Truth and Truthlikeness as Explanations for the Success of Science

Let's have a look at the three extant classes of antirealist arguments against SS. Before we start, it's necessary to do something about the ambiguity in the phrase "A explains B"—and to do it without getting entangled in philosophical questions about explanation that are peripheral to present concerns. To say that A is *an* explanation for B is to make a claim that is consistent with A being false, and with the possibility that some other C, which is incompatible with A, is also an explanation for B. To say that A is *the* explanation for B is clearly to imply that A is true, and that its truth is the reason why B is true. "A explains B" is ambiguous between these two options. I will always use it to mean that A is *an* explanation for B.

Now for the first antirealist sally. According to this argument, neither truth nor truthlikeness can be the explanation for the success of science (I use "truthlikeness" as a synonym for "approximate truth"). This thesis is developed at length by Laudan (1981) in his classic "confutation of convergent realism". According to Laudan, truth fails because the paradigmatically successful scientific theories of the past are now known not to *be* true. Truthlikeness fails for two reasons. First, many the successful theories of the past have not only failed to be true—they've also failed to be approximately true. Second, approximate truth doesn't explain the success of science anyway—it's not even *an* explanation of success, much less *the* explanation.

We'll start with the argument against theoretical truth as the explanation for scientific success. When we're done with this topic, we'll find that most of the issues relating to truthlikeness will have been preempted and that very little additional work is needed to finish up. If the successful theories of the past were false, then it follows either that there is some *other* acceptable explanation for success that applies in those cases or that success sometimes has no acceptable explanation. In either case, the argument from the success of science is refuted by the repudiation of its second premise. Success is apparently not a compelling ground for ascribing theoretical truth. Realists have almost universally abandoned truth as an explanation for success in the face of this argument and have taken refuge in approximate truth. There is a rejoinder available for realists, however. This defense of realism is a minor variant of an argument of McAllister (1993).

McAllister was not, in fact, dealing with the status of truth as an explanans for success. He was trying to defuse Laudan's parallel refutation of the claim that *approximate* truth can be the explanation of success. Laudan maintains that many of the successful theories of the past failed to be even approximately true. Ether theories and phlogistic theories, for instance, posited ontologies that we now completely reject— there is nothing remotely *like* an ether or phlogiston in our current picture of the world. The conclusion, as in the case of truth simpliciter, is that the predictive success of our theories is not a compelling ground for considering them to be approximately true. One realist response to this argument has been to insist that the successful theories of the past *were* approximately true after all (e.g., Hardin & Rosenberg 1982). The problem with this strategy is that the more liberal we make our construal of approximate truth, the more likely it is to succumb to the charge that the approximate truth of our theories doesn't license our taking a realist attitude toward them. If phlogistic theories qualify as approximately true, then a theory may be approximately true even if nothing remotely like its theoretical entities exists. Thus, the truth of premise SS2 is protected by weakening it to the point where it no longer entails the desired conclusion SS3. McAllister takes an alternative line: he admits that the theories cited by Laudan failed to be approximately true, but he also denies that they were successful. To be sure, these theories were *thought* to be successful in their own time, but, according to McAllister, these assessments were made on the basis of criteria for observational success that are now considered to be inadequate. Therefore, Laudan's examples don't attest to his claim that theories far from the truth may nevertheless be successful.

Although McAllister doesn't say so, the same argument can be used to give another lease on life to *truth itself*, as opposed to truthlikeness, as the explanation for success. If it's denied that the theories of the past were ever successful, then the fact that they were all false has no bearing on the truth-based version of SS. What about our best current theories? Let's consider cases. Suppose that even our best current theories aren't successful. The assumption that no successful theory has ever been formulated contradicts premise SS1. Hence, it would undo argument SS. But, as noted in section 2.1, it would not affect the status of the more general success-of-science argument SS', according to which the very possibility of having a successful theory can only be explained by the possibility of having a true theory. This is the place where the assumption that science *isn't* successful is used to bolster the case for realism: it undermines Laudan's argument against the thesis that truth is the explanation of success, without harming the SS' version of the argument for realism.[2]

While my version of McAllister's argument protects SS' from Laudan's confutation, it does so at a cost that realists may be unwilling to pay. They have to assume that quantum mechanics (inter alia) is empirically unsuccessful. This may be an option for those who believe on general skeptical grounds that *no* theories can *ever* be successful. But, as we've already had several occasions to note, realists and antirealists alike must concede that empirical success is at least a possibility—and if *any* theory can be successful, it's hard to understand why quantum mechanics wouldn't fit the bill. What more stringent standards for success could one plausibly adopt than those that quantum mechanics has already passed? So let's assume that our best current theories *are* successful. If these theories are also *false*, then either there are other explanations of success besides truth, or the success of science is inexplicable. In either case, the versions of both SS and SS' that refer to theoretical truth (as opposed to truthlikeness) must fail.

But how compelling is the assumption that our best current theories are false? As far as I know, the only consideration other than private and unarguable intuition that bears on this question is Putnam's (1978) "disastrous induction": all the theories of the past are now known to be false, so it is overwhelmingly likely that our present theories will turn out to be false as well. The premise of the disastrous induction does provide us with *some* negative evidence relating to the hypothesis that our current theories are true. As is the case with all inductions based on bare enumeration, however, the force of its conclusion is easy to resist. It isn't necessary to rehearse how quickly we get into trouble when induction is uninformed by background theories and expectations. Suffice it to say that the falsehood of our past theories, *all by itself*, is a very weak basis for projecting that falsehood onto our present theories. If we were marching steadily toward an attainable goal of absolute theoretical truth, our interim theoretical accounts of the world might nevertheless all be false until we got very close to the end.

In fact, there are background-theoretical considerations that militate against the disastrous induction. Perhaps we should retrace our steps before making this last point about truth. We're considering the realist thesis that theoretical truth is the explanation for theoretical success. Laudan objects to this argument on the ground that the successful theories of the past were *not* true. I counter this with the McAllister shuffle: realists can claim that these false theories weren't really very successful. But the Laudan argument can be run on our current theories as well as our past theories, and it's rather more difficult to suppose that none of our current theories are successful either. On the other hand, it's also not as entirely obvious that our most successful current theories must be considered *false*. The only basis for this opinion is the disastrous induction. Now for the final point: it's been conceded that reasonable realists have to admit that our best *current* theories are successful. But they still don't have to admit that the best theories of the *past* were successful. Moreover, realists are of the opinion that it's success that warrants the inference to truth. Thus, the data cited as the basis for the disastrous induction are completely compatible with realist expectations. If it's accepted that the best theories of the past were unsuccessful whereas the best theories of the present are successful, then realists will rightly conclude that it's inappropriate to project the property of falsehood, which was possessed by the theories of the past, onto the theories of the present. The disastrous induction carries weight only if it's assumed that success is irrelevant to truth. But this assumption begs the question against the realists. And thus the antirealist counterargument against the contention that truth is the explanation of success is a failure.

Now for the pair of counterarguments relating to approximate truth. The realist argument here is that the approximate truth of our theories is the explanation for their success, and that approximate truth entails realism. The first of the two Laudanian counterarguments is that there are successful theories that fail to be even approximately true. We've just seen that antirealists can't nail down the claim that there are successful theories that fail to be *true*. Naturally, the present, stronger claim can only be more problematic. That disposes of the first counterargument.

The second Laudanian counterargument is that there's no reason to believe that approximate truth explains success. Laudan notes that "virtually all proponents of epistemic realism take it as unproblematic that if a theory is approximately true, it deductively follows that the theory is a relatively successful predictor and explainer of observable phenomena" (1984b, 118). This assumption is presumably inherited from "the acknowledged uncontroversial character" of the corresponding claim about truth (118). But the inheritance is by no means automatic. Indeed, it's conceivable that an arbitrarily minute error in our characterization of theoretical entities might result in a drastically incorrect account of observable phenomena. The root problem, as Laudan notes, is that realists have given us no clear conception of how to assess approximate truth. Until they do, it would be philosophical sleight-of-hand to claim that approximate truth is a viable explanation of success.

I accept this criticism of SS. It must be understood, however, that it doesn't constitute a *refutation* of SS—or a confutation, either. What follows from the extreme vagueness of "approximate truth" isn't that approximate truth *doesn't* explain success. It's that it's *hard to tell* whether approximate truth explains success. The proper moral to be drawn from this state of affairs is not that SS is unsound. It's that this particular issue—whether truthlikeness explains success—is not currently a promising arena for a decisive confrontation over the status of SS.

To recapitulate: the Laudanian counterarguments to SS are (1) that success doesn't warrant an inference to truth or truthlikeness because there are successful theories that are neither true nor approximately true, and (2) that there's no reason to believe that truthlikeness is even a formally adequate explanation of success. Neither counterargument is decisive. The first can be defused by insisting on a standard for success that is so high that all the theories whose truthlikeness is in doubt fail the test. If necessary, the standard for success can be raised so high that *all* extant theories fail the test, in which case Laudan's conjunctive existential claim (that there are theories that are both successful and far from the truth) is guaranteed to be false. The second counterargument, while sound, leaves the status of SS unsettled. The fate of SS thus hinges on the results of the other pair of argumentative strategies deployed by antirealists—namely, the attempt to establish the existence of rival antirealist explanations of success, and the attempt to show that realism doesn't follow from the premise that truth is the best explanation of success.

2.3 Antirealist Explanations for the Success of Science

Let's grant that truth and approximate truth explain the success of science. Even so, SS would be derailed if it were shown that there are comparably good *antirealist* explanations of that success. The most thorough investigation of the topic has been undertaken by Leplin (1987, 1997). Laudan and Leplin come to opposite conclusions in their analyses of their respective parts of SS. Whereas Laudan finds no support for realism, Leplin claims that his portion of the argument achieves its aim—that is to say, he concludes that there are no viable antirealist explanations for the success of science. These are not necessarily conflicting results. Leplin may be right in his claim that there are no antirealist rival explanations for scientific success, and Laudan may also be right in maintaining that SS is unsound in other respects. Of course, if both are right, then SS fails. The part of the argument investigated by Leplin would simply not be the part that contains the mistake. In this section, I want to reexamine Leplin's portion of the argument.

Leplin considers two categories of putative antirealist explanations for the success of science: those that allude to evolutionary or kindred mechanisms for theory selection, and those that regard a theory's empirical adequacy already to be an explanation of its success. Leplin argues that both antirealist strategies fail to achieve their purpose, so the only explanation for scientific success on the table is theoretical truth. I am persuaded by his critique of the evolutionary alternatives but not by his critique of the empirical adequacy alternative. I briefly recapitulate the arguments that I agree with, and then turn to the disagreements.

The prototypical evolutionary explanation for the success of science is van Fraassen's (1980), according to which the success of our scientific theories is due to their having been winnowed through a process of Darwinian selection:

> Species which did not cope with their enemies no longer exist. That is why there are only ones who do. In just the same way, I claim that the success of current scientific theories is no miracle. It is not even surprising to the scientific (Darwinist) mind. For any scientific theory is born into a life of fierce competition, a jungle red in tooth and claw. Only the successful theories survive—the ones which *in fact* latched on to actual regularities in nature. (39)

The doctrine espoused by van Fraassen—that theories are selected by the scientific community for predictive success and that this practice suffices to account for success—often goes by the name of "evolutionary epistemology". A related account, confusingly also called "evolutionary epistemology", has it that it's the evolution of human beings that explains the success of their science. On this view, scientific theories are successful because their inventors are the outcome of a process of natural selection that favored beings who abduce successful theories. To a certain extent, these two evolutionary epistemologies require different critiques. For instance, van Fraassen's version has to be defended against the charge that the competition among theories fails to satisfy some of the conditions necessary for a process of Darwinian selection to take place. Let's grant, for the sake of the argument, that either of these accounts is true. Even so, Leplin's (1997) analysis shows that either account fails to provide an explanatory alternative to theoretical truth. I'll summarize his discussion of van Fraassen's account. The same story also undoes the biological version of the argument.

The problem, according to Leplin, is that the question answered by evolutionary epistemology is not the same as the question answered by theoretical truth. The tooth-and-claw account explains how we come to be in *possession* of successful theories, but it doesn't explain why it is that *these* theories, which happen to be the ones we possess, are successful. Leplin draws an analogy between the success of science and the tennis-playing excellence of Wimbleton finalists. How do we explain the fact that Wimbleton finalists are so good at tennis? This question is amenable to two interpretations. If we want to know why Wimbleton finalists in general are good tennis players, it's appropriate to cite the stringency of the selection procedures for entry into the tournament. However, this explanation doesn't tell us why *these particular individuals*, Borg and MacEnroe, who happen to be Wimbleton finalists, are so great at tennis. To answer this second question, we would have to cite their own relevant properties, such as their training and genetic attributes. Now theoretical truth answers the second type of question in relation to predictive success. It cites an attribute of *these particular theories* that accounts for why they're successful. It does not, however, explain, or even try to explain, how we come to possess these successful theories. The question "why are the theories that we possess, whatever they may be, successful?" doesn't even receive *an* explanation, much less the correct explanation, by the answer "because they are true". Conversely, the question "why is quantum mechanics successful?" is given at least *an* explanation by saying "because it's true", but it isn't answered at all by saying "because its adoption is the result of a process of Darwinian selection for success". In sum, evolutionary epistemology and theoretical truth are not explanatory rivals.

Leplin notes that the same can be said about Laudan's (1984a) explanation of the success of science. According to Laudan, success is explained by the fact that scientists use appropriate experimental controls such as double-blind methodologies in testing their theories. Clearly, this proposal belongs to the same class as van Fraassen's. In fact, it can be regarded as an elaboration of the tooth-and-claw story: van Fraassen tells us that the success of our scientific theories is due to their having been winnowed by a process of selection, and Laudan specifies the nature of the selective mechanism. Once again, the fact that our theories have been tested by such-and-such a methodology may explain why we possess successful theories. But it doesn't explain why quantum mechanics and not some other theory is the one that has survived all the tests. The hypothesis that quantum mechanics is true, however, provides at least an explanation for its success.

Now for the disagreements. Before launching into the topic, I wish to set a few ground rules for the discussion. Naturally, it will be supposed for the sake of the argument that none of the other potential problems with SS need to be worried about. I assume that science *is* successful, that argument SS is *not* question-begging or invalid, that approximate truth *is* both meaningful and measurable, and that approximate truth *does* explain the success of science. Given these powerful assumptions, it makes no difference whether we talk about the truth of a theory or its truthlikeness. I will therefore be cavalier in my treatment of this distinction, as well as with the distinction between empirical adequacy and approximate empirical adequacy. For the sake of expository convenience, I'll usually speak of the truth of theories when I mean their truth or truthlikeness, and of the empirical adequacy of theories when I mean their empirical adequacy or approximate empirical adequacy. I provisionally define "T is (approximately) empirically adequate" as "T is (approximately) empirically equivalent to a true theory". The reason for this eccentricity will become apparent below.

Leplin (1987) coins the term *surrealism* for the view that the explanation for the success of our theories is that they are empirically adequate in the eccentric sense in which I've provisionally defined this term — that is, that our theories are empirically equivalent to true theories. Surrealism holds that the observable world behaves *as if* our theories were true. But it "makes no commitment as to the actual deep structure of the world". Surrealism "allows that the world has a deep structure" — that is to say, it allows that there is a true theory of the world whose theoretical terms denote real entities. But it "declines to represent" these structures (Leplin 1987, 520). Leplin attributes the development of this view to Arthur Fine (1986). As Leplin notes, Fine doesn't actually endorse surrealism. He merely cites its coherence to discredit the miracle argument.

Leplin presents two arguments to the effect that surrealism collapses into realism. The first is that the surrealist explanation for the success of science *presupposes* the realist explanation:

> To suppose that Surrealism explains success is to suppose that for the world to behave as if the theory is true *is* for the theory's predictions to be borne out. But this is to suppose that theoretical truth is normally manifested at the observational level, and thus to *presuppose* the original, realist explanation of success rather than to provide a weaker alternative. (1987, 523)

If this is right, then surrealism doesn't qualify as an *alternative* to the realist explanation. It seems to me, however, that this claim merely equivocates between the two senses in which one thing may be said to explain another. To suppose that the empirical adequacy of a theory T explains its success (in either sense of "explains") is indeed to presuppose that T's truth must also qualify as *an* explanation of its success. But it is not to presuppose that the truth of T is the *correct* explanation of its success or even that T is true at all. After all, realists and antirealists alike acknowledge that scientists frequently make use of theories that are known to be false in predicting experimental results. For example, scientists sometimes represent gas molecules as tiny billiard balls. The explanation for why these patently false theories succeed is presumably surrealistic: they're approximately empirically equivalent to other theories about gases that are true. If Leplin's claim that surrealism presupposes realism were to be accepted, we would have to conclude that gas molecules *are* tiny billiard balls.

It might be argued that, while approximate empirical adequacy doesn't presuppose truth simpliciter, it does presuppose approximate truth: the success of the billiard ball theory of gases may not entail that gas molecules are billiard balls, but it does entail that gas molecules are rather *like* billiard balls. Moreover, most realists would be content with the verdict that our best theories are approximately true. Thus, the surrealist explanation of the success of science may presuppose the realist explanation after all. Does approximate empirical adequacy entail approximate truth? The temptation to think so stems from the following considerations. For a theory to be approximately true means (let us say) that it gets many of its claims right, even though it may also get some things wrong. By the same token, for a theory to be approximately empirically adequate means that it gets a lot of its *observational* claims right, even though it may get some of them wrong. But to get a lot of observational claims right is to get a lot of claims right *tout court*. Therefore, approximate empirical adequacy entails approximate truth. The problem for realists is that this notion of approximate truth has no connection whatever to scientific realism. If getting *any* sizable part of the story right is already sufficient for approximate truth, then it's possible for approximately true theories to get the *theoretical* part of the story *completely* wrong. Getting a lot right is certainly *necessary* for approximate truth. But if approximate truth is to have any bearing on the realism issue, it must also be stipulated that approximately true theories get at least some things about theoretical entities right. There's no need here to specify just what part or how much of the theoretical story an "approximately true" theory has to get right. But however slight the theoretical requirement may be, the result is a notion of approximate truth that is no longer presupposed by approximate empirical adequacy: getting an arbitrarily large part of the observational story correct is logically compatible with getting the theoretical story totally wrong in every respect. And thus the surrealist explanation doesn't presuppose the realist explanation.

Leplin's second argument is that surrealism threatens to collapse into realism unless every theory is guaranteed to have empirically equivalent rivals (let EE be the hypothesis that every theory does have empirically equivalent rivals). For suppose that we have a successful theory T that has no empirically equivalent rival. Then the surrealist explanation for T's success—that T is empirically equivalent to a true theory—entails that T itself is true. That is to say, the surrealist explanation entails the realist explanation. To maintain its distinctiveness, surrealism must therefore be *committed* to the truth of EE. Now, Leplin regards it as an open question whether EE is true, hence as unproved that surrealism provides a genuine explanatory alternative. My first response to this argument is that the universal availability of empirically equivalent rivals is not an open question at all. In chapter 5, I argue that there are several algorithms for transforming any theory into an empirically equivalent rival theory. I won't insist on this point here, however. Let us for the sake of the argument agree that the status of EE is uncertain, hence that the distinctiveness of the surrealist explanation of success is also uncertain. I will try to show that surrealism can be weakened in a manner that blocks the derivation of realism from the negation of EE but that still leaves us with an explanation for the success of science.

At this point, let's revert to the more traditional definition of empirical adequacy: T is empirically adequate if all of its empirical consequences are true. This notion is logically weaker than the previous one, by virtue of the fact that it drops the requirement that T's empirical consequences also be the consequences of a true theory. Just as before, the empirical adequacy of T entails that the observable world behaves as if the "deep structures" postulated by T exist, but it's no longer required that the observable phenomena be consequences of any real "deep structures". Let's give the names *weak surrealism* to the view that our best theories are empirically adequate in the new diminished sense and *strong* surrealism to the thesis that our best theories are empirically equivalent to true theories. Now consider the hypothesis that weak surrealism explains the success of science.

Whatever its other shortcomings may be, this hypothesis is immune to the second difficulty that Leplin makes for the strong surrealist explanation. However, Leplin claims that this escape from the collapse of strong surrealism merely lands us in another dilemma:

> The statement that things happen as they would were our theories true invites two directions of analysis. On one analysis, Surrealism simply tells us what goes on at the observational level. Then it is not simply metaphysically weaker than realism; it is not metaphysical at all. But neither is it explanatory. Rather, it restates its explanandum. (1987, 522)

The "direction of analysis" alluded to in this passage is the direction of weak surrealism. The other direction is strong surrealism. The way Leplin sees it, in their choice between strong and weak surrealism, antirealists are caught between the Scylla of realism and the Charybdis of vacuity. If they opt for strong surrealism, their explanation threatens to become indistinguishable from the realists'. But if they choose weak surrealism, then their explanatory hypothesis amounts to nothing more than a recapitulation of the explanandum.

My criticism of this argument is that it conflates three putative explanations of success into two: there is a position between strong surrealism and the explanation that Leplin regards as vacuous. Let T* be the set of all the empirical consequences of T. Weak surrealism is the view that T* is true. But there are two different ways of holding that T* is true. On the one hand, one may believe that "the observable world behaves as if T is true" is itself a *nomological* truth about the world. On the other hand, one may believe in the truth of a set X of nomological empirical generalizations, as well as in the *non*nomological hypothesis that X is coextensive with T*. The difference between these two cognitive states is brought out by asking what happens if some previously unexamined empirical hypothesis E is found to be a consequence of T. If you believe that it's a law of nature that the world behaves as if T is true, then you would add E to your stock of beliefs. But if you think that the coextensiveness of T* and your empirical beliefs is merely accidental, then the same discovery would leave the status of E entirely open: its derivation from T would not be a rationally compelling reason for its adoption. Indeed, it might be a good reason for giving up the belief that the true empirical generalizations are coextensive with T*. Let's reserve the name "weak surrealism" for the first view and call the second view *fragmentalism*. Clearly, fragmentalism is logically weaker than weak surrealism.

Now consider the following series of progressively weakened views about theories: (1) realism (our theories are true), (2) strong surrealism (our theories are empirically equivalent to true theories), (3) weak surrealism (the observable world behaves as if our theories are true), and (4) fragmentalism (the true empirical generalizations are coextensive with the empirical consequences of our theories). The further down the list we go, the less danger there is of a collapse into realism, but also the fewer resources we have for explaining the success of science. Leplin concedes that strong surrealism explains the success of science but claims that there's serious danger of its collapsing to realism. On the other hand, a fragmentalist explanation of success falls prey to Leplin's charge of vacuity. The fragmentalist believes that the ultimate truth about the world—or as much of it as we can ascertain—is given by a collection of empirical generalizations. But the truth of our empirical generalizations is exactly the explanandum in the success of science argument. Leplin doesn't distinguish the vacuous fourth explanation from the third. But the third—weak surrealism—seems to avoid both Scylla and Charybdis. According to weak surrealism, the truth of our empirical generalizations is explained by the facts that (1) these generalizations are consequences of T, and (2) the world behaves as if T were true. Since this claim goes beyond what is contained in the explanandum, weak surrealism avoids the charge of vacuity. At the same time, there's no danger of this view collapsing to realism, even if it turns out that some theories have no empirically equivalent rivals. Thus, weak surrealism is an antirealist rival to the realist explanation of the success of science.

Of course, to concede that there are coherent and nonvacuous antirealist explanations for success is not yet to admit that these explanations are as good as theoretical truth. What about the weaker claim that the realist explanation is better than the antirealist explanation? One can see what the realist's intuition is here: the weak surrealist claim that the empirical adequacy of T is a fundamental fact about the world strikes them—and me—as *unintelligible*. We have no idea what could possibly keep all the disparate empirical consequences of T in line, if not the causal mechanisms posited by T itself, or at least, as strong surrealism would have it, the mechanisms posited by some other theory that is empirically equivalent to T. Simply saying that they *are* kept in line doesn't explain anything. I grant the unintelligibility of weak surrealism, but I question the epistemic import of intelligibility. I suspect that the intelligibility of explanations is culturally relative. It used to be thought that teleological explanations give us understanding. Now we want to know the mechanism behind the appearance of teleology. More to the point, explanations that posit action at a distance have been considered intelligible and unintelligible in turn by successive generations. One could certainly express the same puzzlement over action at a distance as one does over surrealism: we have no idea how action is propagated across a void, and just saying that it happens doesn't explain anything.

One could say, so much the worse for action at a distance. But where does the intelligibility of the *acceptable* explanatory mechanisms come from? Presumably, realists like Leplin have in mind a paradigm of explanations that satisfy our explanatory hunger. Maybe the most intelligible kind of explanation we can have is a narrowly "mechanistic" one that explains phenomena in terms of particles pushing each other around. When we can account for something by means of this sort of mechanism, then we feel we have understood it as well as we can understand anything. But does the intelligibility of this type of mechanistic explanation consist of anything more than a psychological willingness to stop inquiry at that point? After all, how does "pushing around" work? When A makes contact with B, why don't they simply interpenetrate without interacting? Just to say that they *do* push each other around doesn't explain anything. It is, I grant, premature to claim that there *cannot be* an epistemologically relevant notion of intelligibility that brands surrealism as unintelligible. But

I've encountered nothing in the philosophy of science literature to suggest that there *is* such a notion. On the contrary, the only relevant discussion that I know of takes the view that the intelligibility of scientific theories is a matter of "psychological acclimation" (Cushing 1991, 346).

Realists might complain that the foregoing critique breaks one of the ground rules for this discussion. It was supposed to be assumed that all of the other qualms about SS had been allayed. One of these was the objection that SS is invalid. After laying down the ground rules, I claimed, contra Leplin, that weak surrealism is a coherent, distinct, and nonvacuous explanation for the success of science. If I'm right, then realists need to show that theirs is a *better* explanation than the weak surrealists'. They do not, however, have to show that explanatory goodness has epistemic import. This principle was already tacitly assumed when it was granted that SS is valid—for if explanatory goodness were *not* epistemically relevant, then the conclusion of SS wouldn't follow from its premises. Yet isn't it precisely this principle that I call into question when I argue that the intelligibility of explanations may be culturally relative? I suppose it is. But to grant that explanatory goodness is epistemically relevant is not yet to give realists the license to use any features of explanations they like in making a comparative evaluation. It would clearly be cheating, for instance, if realists declared their explanation of success to be superior to the weak surrealist explanation on the grounds that *the positing of theoretical entities* is an indicator of explanatory goodness.

The problem is that the general assumption that explanatory goodness is epistemically relevant doesn't yet tell us which theoretical properties are indicative of this epistemically relevant explanatory virtue. When it comes to this phase of the argument, we can no longer deal with Leplin's problem in isolation from the other problems faced by SS. The task for realists is to find a property of explanations that simultaneously satisfies this pair of criteria: (1) theoretical truth has more of it than empirical adequacy, and (2) its possession is relevant to beliefworthiness. At a minimum, realists need to show that there's a property that theoretical truth has more of than empirical adequacy and that is not demonstrably *ir*relevant to beliefworthiness. It's trivially easy—and unilluminating—to satisfy either requirement by itself. Theoretical truth is undoubtedly a more intelligible explanation of success than empirical adequacy. But if it's true that intelligibility is culturally relative, then it can't play the requisite role in any epistemology that is compatible with realism. Conversely, an example of a property of explanations whose epistemic credentials can't be denied is *the probability that it is true*. But since theoretical truth is logically stronger than empirical adequacy, it can't coherently be maintained that the realist explanation has more of this property than the surrealist explanation. The trick is to satisfy both requirements at once. Realists haven't yet managed to perform this trick.

2.4 The Circularity of the Argument from the Success of Science

Let's take stock of where things stand with SS. In section 2.2, I argued that Laudan's demonstration that neither truth nor truthlikeness can explain the success of science is inconclusive. In section 2.3, my point was that Leplin's demonstration that there are no antirealist explanations for scientific success likewise fails to be conclusive. By my reckoning, neither side in the debate can yet be considered the victor. But the most powerful antirealist counterargument is still to come. The charge, made independently by Laudan (1981) in his confutation and by Fine (1984), is that SS is circular. Here's how the argument goes.

Grant that science succeeds, and that the truth of our scientific theories is the *only* viable explanation for that success. Even so, the conclusion of SS—that we have grounds for believing our scientific theories—doesn't follow unless it's assumed that the explanatory virtues of hypotheses are reasons for believing them. But this is an assumption that the antirealist need not accept. In fact, antirealists generally don't accept it. Van Fraassen (1980), for instance, distinguishes between the *epistemic* and *pragmatic* virtues of theories. Both are features that we would like our theories to have, but only the former bear on their beliefworthiness. It's obvious that there are many—perhaps indefinitely many—pragmatic virtues. One of them is the property of allowing us to make calculations quickly and easily. Suppose T1 and T2 are empirically equivalent and that calculating empirical predictions is easier with T1. This is an excellent reason for *using* T1 instead of T2 in deriving our predictions. But it's not necessarily a reason for thinking that T1 is closer to the truth. Most of us would want to say that ease of calculation is merely a pragmatic virtue.

It's van Fraassen's view that the only epistemic virtues of theories are the *empirical* virtues of getting more observable consequences right or fewer of them wrong. An immediate corollary of this belief is that the *explanatory* virtues of theories, being other than empirical, are merely pragmatic. The fact that a theory provides the best— or even the only—explanation of a set of phenomena thus has no bearing on its beliefworthiness. If antirealists are right on this score, then SS fails even if both of its premises are true. Of course, antirealists may be mistaken about epistemic virtues. I investigate the force of their hypothesis about epistemic virtues in chapter 6. But right or wrong, it's question-begging to wield an argument against them that merely *presumes* that explanatoriness is a reason for belief. But this is exactly what is presumed in SS. Thus, the argument from the success of science accomplishes nothing in the debate between realists and antirealists.

Laudan's and Fine's circularity counterargument has been responded to by Boyd (1984). Boyd claims that scientists routinely use "abductive inference" in choosing between *empirical* hypotheses—that is, they decide what to believe about the observable world on the basis of which hypothesis best explains the data. But then, Boyd argues, it must also be permissible for philosophers to rely on abduction in defense of a philosophical thesis about science. This move has been countered in turn by Sober (1990), who claims that the problem with SS isn't that its use of abduction is question-begging, but rather that it's a very *weak* abductive argument. I think that the original charge of circularity can be sustained against Boyd's objection. Boyd's defense belongs to a recurring pattern of realist argumentation that I criticize in detail in chapter 6. At present, a quick overview of my objection will do. Realist arguments belonging to this class note that antirealists refuse to give epistemic weight to some nonempirical theoretical virtue like simplicity or explanatoriness when these apply to theoretical statements, but that they're willing to use the very same principle when dealing with observational claims. This practice is supposed to convict the antirealist of internal inconsistency. But the evidence doesn't warrant such a conviction. It's simply not true that the appeal to nonempirical virtues in assessing the status of observational hypotheses logically commits us to applying the same principles to theoretical hypotheses. Suppose we use some rule R for giving epistemic weight to, say, the explanatory virtues of hypotheses: R tells us to give greater credence to hypotheses on the basis of how much and how well they explain. This rule may or may not specify circumstances under which we should elevate the epistemic status of theoretical hypotheses. In case it does, let R* be the same rule with the added proviso that it applies only to observational hypotheses. Antirealists commit no *logical* impropriety in subscribing to R*. Yet R* allows for abductions to observational hypotheses while blocking abductions to theoretical hypotheses. To be sure, this account of the matter leaves antirealism open to the lesser charge of arbitrariness. But arbitrariness is not, by itself, a decisive counterargument against a philosophical position. Indeed, it can be a legitimate part of a philosophical position, as is the arbitrariness of prior probability assignments in personalism. I have more to say about arbitrariness in several of the later chapters. For the present, I note only that it isn't a decisive objection. Thus, a small change in Laudan's and Fine's argument insulates that argument from Boyd's critique. Laudan and Fine say that the use of abduction in SS is question-begging, since antirealists deny the validity of abduction. Boyd counters that everybody uses abduction, including antirealists. This may be so. But antirealists certainly will not allow the use of abduction in the service of theoretical hypotheses. But realism itself is a theoretical hypothesis, for it entails that some theoretical entities exist. Thus, endorsers of SS are guilty of begging the question by engaging in *abduction to theoretical hypotheses*, when such abductions are just what antirealists regard as illegitimate.

I have more to say about SS in chapter 3. These further remarks, however, merely heap additional criticisms on the argument. The conclusion of the present section stands: the success of science argument fails to do the job it was designed for.

Realism and
Scientific Practice

In the debate about scientific realism, both sides have made the claim that their position provides a better (or the only) account of some (or all) features of actual scientific practice. On the realist side, the most famous of this class of claims is undoubtedly Putnam's (1975b) conjunction argument, which runs as follows. Suppose that scientists accept theories T1 and T2, and that the empirical hypothesis E is a consequence of the conjunction T1 & T2, but not of either T1 or T2 in isolation. In this situation, it is claimed, scientists are willing to *believe* that E is true. This inference is quite understandable under the realist hypothesis that acceptance of T1 and T2 is a matter of believing that these theories are true — for if we believe both T1 and T2, then it's rational also to believe any consequence of their conjunction. But the same inferential step would be a simple fallacy if scientists were in fact instrumentalists or van Fraassian constructive empiricists, in which case their acceptance of T1 and T2 would merely involve a belief in the empirical adequacy of T1 and T2 — for there's no guarantee that the conjunction of two empirically adequate theories is itself empirically adequate. Thus, we must either accuse science of irrationality or admit that realism is true.

An argument from scientific practice on the *anti*realist side has been presented by Morrison (1990). Following Cartwright (1983), she notes that scientists routinely employ multiple and incompatible theories in making empirical predictions. If theories are given a realistic interpretation, this practice would be blatantly irrational. Therefore theories should not be interpreted realistically. Morrison cites the multiple modeling of the behavior of gases as an example:

> The overall difficulty seems to be one of specifying a molecular model and an equation of state that can accurately and literally describe the behavior of gases. We use different representations for different purposes; the billiard ball model is used for deriving the perfect gas law, the weakly attracting rigid sphere model for the van der Waals equation and a model representing molecules as point centers of inverse power repulsion is used for facilitating transport equations. What the examples illustrate is that in order to achieve reasonably successful results we must

vary the properties of the model in a way that precludes the kind of literal account that [realists prescribe]. . . . Instead an explanation of the behavior of real gases . . . requires many *different* laws and *incompatible* models. (1990, 319)

The two arguments have a very similar structure. In both cases, it's claimed that a certain methodological practice is prevalent among scientists, and that the rationality of this practice presupposes a broad philosophical thesis. The problem is that the two theses are mutually inconsistent. The conclusions can't both be right. Section 3.1 is devoted to finding the mistake. There are various possible modes of resolution. One possibility is that one or the other side in the debate is wrong about what it considers to be a scientific practice. Van Fraassen (1980), for instance, tries to disarm the Putnam argument by contending that scientists never do conjoin theories in the way that Putnam describes. Another possibility is that, despite the appearance of symmetry, one side has a better case for its conclusion than the other. I reject both these alternatives. I claim that both sides are right about the prevalence of the scientific practices they allude to, and that their arguments are indeed symmetric. What accounts for their incompatible conclusions is that neither argument can be made to work.

There's another way to account for the disparate conclusions of realists and antirealists about scientific practice, which I won't explore here. It may be that both sides are right after all—that the practices they both allude to are bona fide scientific practices and that these practice do presuppose their respective doctrines. What would account for such a state of affairs is the hypothesis that science is irrational—that some of its practices presuppose realism and others presuppose antirealism. There are, of course, more than a few philosophers of science who regard scientific practice as *constitutive* of rationality. This is not my view. But the subject is too broad to bring to bear on the issue at hand. I'm willing to accept the rationality of science for the purpose of the present discussion, since my case for claiming that neither the realist's nor the antirealist's arguments succeeds doesn't depend on the *ir*rationality of any putative scientific practice.

The refutation in section 3.1 of the conjunction and multiple models arguments leaves open the possibility that other scientific practices may yet be shown to presuppose realism or antirealism. In section 3.2, I consider a general argument of Arthur Fine's (1986) to the effect that antirealists can explain any practice that realists can explain. I claim that this argument falls short of achieving its aim. However, I agree with Fine that nobody has ever demonstrated an incompatibility between any scientific practice and either realism or antirealism.

In section 3.3, I take up an argument of van Fraassen's (1980, 1985) that concedes, or at least doesn't presuppose the negation of, the hypothesis that both realists and antirealists can account for all scientific practices. Van Fraassen maintains that the antirealist account of scientific practice is in every instance weaker, hence better than its realist competitor. I claim that this argument begs the question against the realist and that, moreover, even if the argument were sound, it would not accomplish anything useful for the antirealist's case. The overall conclusion of the first three sections is that, contrary to the opinions of many realists and antirealists alike, the facts about scientific practice have not been shown to have any bearing on the realism issue.

Sections 3.4 and 3.5 contain what I take to be important addenda to the foregoing analysis. In section 3.4, I argue that the structure of van Fraassen's argument from scientific practice is precisely mirrored in several other famous arguments both for and against realism, and that these further arguments suffer from the same pair of shortcomings: circularity and redundancy. My pessimistic assessment of the state of the debate is reminiscent of Fine (1984), whose response to this dilemma has been that we should eschew both realism and antirealism and instead adopt the "natural ontological attitude". In section 3.5, I show that the argument for the natural ontological attitude involves the same question-begging maneuver as van Fraassen's argument and its several analogues.

3.1 The Failure of the Theory Conjunction and Multiple Models Arguments

Let's begin by asking whether the practices alluded to in these arguments are bona fide scientific practices. As far as I know, no one has ever tried to deny that scientists employ multiple models. However, van Fraassen (1980) has famously denied that scientists engage in theory conjunction. Morrison (1990) has made the same point. Van Fraassen's reply to the conjunction argument, in brief, is that the history of science does not show that the conjunction of two accepted theories is itself accepted "without a second thought" (1980, 85). In fact, there "can be no phenomenon of the scientific life of which this simple account draws a faithful picture" (85). According to van Fraassen, theories are never simply conjoined without further ado:

> In practice scientists must be very careful about conjoining theories, . . . because . . . it rarely happens that non-tentative, unqualified acceptance is warranted. . . . Putting . . . two theories together would not [mean] conjunction, but correction. (1980, 83–84)

What follows from *this* observation, in turn, is that the actual scientific practice is adequately accounted for by an antirealist stance such as van Fraassen's constructive empiricism. The process of "correction" that van Fraassen refers to is certainly commonplace in the history of science. In particular, as Morrison (1990) notes, historical instances of theoretical *unification* in science have not proceeded by a straightforward logical conjunction of two prior theories. Nevertheless, van Fraassen is wrong in supposing that there is no phenomenon in the scientific life that involves a straightforward conjunction. Trout (1992) has pointed out that just such a conjunction takes place every time scientists use a theory as an auxiliary hypothesis in deriving empirical consequences from the theory they're working on. Contrary to van Fraassen's claim, this "mercenary" use of theoretical auxiliaries doesn't involve any corrections of the auxiliary theory. Indeed, the user is typically inexpert in the field that the auxiliary comes from and is thus unqualified to suggest theoretical revisions. So Putnam's argument can't be dismissed on the grounds that its premise about scientific practice is false.

What *is* wrong with the conjunction argument, however, is its supposition that an antirealist can have no warrant for conjoining theories. If "constructive empiricism" is taken to refer to the totality of van Fraassen's views on the philosophy of science, then I grant that constructive empiricists can't allow themselves to conjoin theories. But there are other forms of antirealism that can. For any theory T, let T* be the hypothesis that all the empirical consequences of T are true. Constructive empiricism is the view that we're never warranted in believing more than T*. This view is susceptible to Putnam's conjunction argument because it's a fallacy to infer $(T_1 \& T_2)^*$ from T_1^* and T_2^*. Now, for any theory T, let T# be the hypothesis that the empirical consequences of T, *in conjunction with whatever auxiliary theories are accepted*, are true. Belief in T# is stronger than belief in T*. Thus, believers in T# are more liberal than van Fraassian constructive empiricists who would restrict our circle of beliefs to T*. But belief in T# is not yet the full belief in T recommended by realists. In particular, the truth of T# doesn't entail that the theoretical entities posited by T exist. That is to say, believers in T# are still antirealists. Let's call believers in T# by the name of *conjunctive* empiricists.

Here is a more precise characterization of conjunctive empiricism. To be a conjunctive empiricist is to specify a set of theories — those that are "accepted" — such that one believes in the empirical consequences of any conjunction of theories in the set. The assertion "T#" means that T belongs to this set. Like their constructive empiricist cousins, conjunctive empiricists will only allow themselves to believe that the world behaves *as if* T were true. But where constructive empiricists interpret this to mean only that the empirical consequences of T are true, conjunctive empiricists go a step further: they think that phenomena will also confirm all the empirical consequences that follow from the conjunction of T with other accepted theories.

It might be objected that the supposition that T# could be true without T also being true is utterly fantastic. But this objection does nothing more than reiterate Putnam's (1975a) other famous argument for realism — the miracle argument. In the original miracle argument, the claim is that the truth of T explains the otherwise "miraculous" truth of T*; here the argument is that the truth of T explains the even more miraculous truth of T#. It was seen in chapter 2 that the arguments of Fine (1984) and Laudan (1981) show the original miracle argument to be circular. Obviously, these arguments apply with equal force to the present version.

The point of introducing conjunctive empiricism is this: conjunctive empiricists have just as good a reason to conjoin theories as realists have. It is indeed a fallacy to infer $(T_1 \& T_2)^*$ from $(T_1)^*$ and $(T_2)^*$. But the inference of $(T_1 \& T_2)\#$ from $(T_1)\#$ and $(T_2)\#$ is unobjectionable. For suppose that $(T_1)\#$ and $(T_2)\#$ are true — that is, that T_1 and T_2 are "accepted" in the conjunctive-empiricist sense of the word. Then any empirical consequence of T_1 (or of T_2) in conjunction with any other accepted theories is true. To show that $(T_1 \& T_2)\#$ is true, it has to be established that any empirical consequence of $T_1 \& T_2$ in conjunction with any other accepted theory is true. Suppose that E is such a consequence and that E is obtained from the conjunction of $T_1 \& T_2$ with the accepted theory T_3 — that is to say, E follows from $T_1 \& T_2 \& T_3$. Now, all three of these theories — T_1, T_2, and T_3 — are hypothesized to be accepted theories. Thus, it can also be said that E is an empirical consequence of the conjunction of T_1 and other accepted theories. But *that's* just another way of saying that E follows from $(T_1)\#$. Since $(T_1)\#$ is assumed to be true, it follows that E is true. This completes the proof that (T1)# and (T2)# entails (T1 & T2)#, which establishes that the practice of theory conjunction doesn't presuppose realism.

It's worth noting that the foregoing argument concedes that considerations of scientific practice do have *some* epistemic import. In particular, if it's granted that science is rational, then the practice of theory conjunction does seem to militate against van Fraassian constructive empiricism. The thesis I wish to defend is that the contemplation of scientific practices hasn't helped to settle the broad issue that divides realists from antirealists.

Let's turn now to the argument for antirealism from the practice of using multiple models. As noted earlier, nobody has ever tried to deny that scientists employ multiple and incompatible models. However, the argument fails because realists can also make empirical predictions from multiple incompatible models without being irrational. Morrison (1990) considers the following realist account of the use of multiple models, which she attributes to an anonymous referee: there is a true theory of the phenomena we are interested in, but this theory is yet to be discovered. In the meantime, we can use the various incompatible theories that have already been formulated, as long as we believe them to be empirically adequate in a specifiable domain. Morrison's reply is that the use of multiple models is virtually universal. Thus, if we use the anonymous referee's defense, "there is very little, if anything, that current theory tells us which can be literally understood" (322). Even if Morrison is right on this point, it leaves unchallenged the referee's contention that realism is compatible with the use of multiple models. Furthermore, realists can rationally avail themselves of multiple models even if they think that they already possess the true theory of the phenomena. For example, they may predict an experimental result by using an alternative theory that they believe to be false, as long as they have reason to believe that the false theory's prediction is sufficiently close to the true theory's over a specifiable domain of phenomena. They might choose to do so because the analysis from the true theory is prohibitively complex. Realists who believe in quantum mechanics are not thereby barred from tackling certain classes of problems with the apparatus of classical mechanics.

Thus, the conjunction argument for realism and the multiple models argument for antirealism fail for the same reason: both practices are available to rational adherents of either philosophy.

3.2 The Unavailability of Any Other Incompatibility Arguments

The analysis of section 3.1 leaves open the possibility that a detailed study of various scientific practices may yet turn up something else that is incompatible with either realism or antirealism. Let's look at each of these two possibilities in turn. First, what are the prospects of finding a practice that is compatible with realism but not with antirealism? Fine (1986) claims to have a general argument which shows that the prospects are absolutely nil. He presents a universal algorithm for turning any realist account of a phenomenon into a corresponding antirealist account. If the algorithm works, we can be sure on a priori grounds that no amount of searching through the annals of science will ever turn up a practice that realists can live with but antirealists can't. The argument runs as follows:

> The realist must at least allow that, generally speaking, truth does lead to reliability. For instance, in the context of the explanationist defence, the realist offers the truth of a theoretical story in order to explain its success at a certain range of tasks. If this offering is any good at all, the realist must then allow for some intermediate connection between the truth of the theory and success in its practice. The intermediary here is precisely the pragmatist's reliability. Let us therefore replace the realist's 'truth' with the pragmatic conception, framed appropriately in terms of reliability. Then, if the realist has given a good explanatory account to begin with, we get from this pragmatic substitution a good instrumentalist account of the same phenomena. Indeed, the instrumentalist account we obtain in this way is precisely the intermediate account that the realist has himself been using. Since no further work is done by ascending from that intermediary to the realist's 'truth', the instrumental explanation has to be counted as better than the realist one. (154)

Fine maintains that the foregoing argument proves what he calls "Metatheorem 1":

> If the phenomena to be explained are not realist-laden, then to every good realist explanation there corresponds a better instrumentalist one. (154)

Thus, consider the miracle argument for realism. According to Putnam (1975a), the truth of our theories provides us with an explanation for their predictive success. By applying Fine's algorithm, we obtain a better antirealist account of the same phenomenon, namely, that it's the "instrumental reliability" — roughly, the empirical adequacy — of our theories that accounts for their predictive success. This is the thesis that Leplin calls "surrealism" (see section 2.3). But Metatheorem 1 has implications that go beyond surrealism. It also entails, inter alia, that from every realist account of a *scientific practice*, we can construct a corresponding antirealist account of the same practice by substituting "empirical adequacy" for "truth". Take the practice of theory conjunction. The realist claims that conjunction makes sense only if we believe that the theories conjoined are true. The result of applying Fine's algorithm to this account is precisely what I called "conjunctive empiricism": the practice of conjunction makes sense under the less-than-realistic belief that the conjunction of any two accepted theories is empirically adequate. In fact, Fine would say that conjunctive empiricism provides a *better* account of the practice of conjunction than realism does.

Let's postpone consideration of whether the corresponding antirealist explanation is actually better than its realist counterpart. Let us first ask whether it's true that there always *is* an antirealist rival to every realist account. In fact, I want to restrict the discussion to the question of whether there exists an antirealist account corresponding to every realist account *of a scientific practice*. Let's call this special case of Metatheorem 1 by the name *Metalemma 1*. If Fine's argument for Metalemma 1 is sound, then the substitution of "empirical adequacy" for "truth" will turn any realist account of a scientific practice into a competing antirealist account. I don't think that the matter is as simple as that. The algorithm works when the role played by truth in a realist account is indeed mediated by empirical adequacy. This is the case in both the miracle argument and the conjunction argument. But there are also cases where this crucial prerequisite for the application of Fine's algorithm fails to be met.

Suppose, for instance, that the accepted theory T of some domain posits a number of laws involving several theoretical entities. Realists regard the empirical success of this theory as evidence for the existence of its theoretical entities; antirealists don't. Now suppose that a theoretician has a hunch that one of these laws, L, can be eliminated from the theory without altering the theory's empirical consequences. Furthermore, the elimination of L would expunge all references to one of T's theoretical entities. As things presently stand, L is in fact used in some important derivations of empirical consequences from T. But the theoretician suspects that there are alternative derivations of the same phenomena that are more complex but that don't require L. Is this theoretician's hunch worth pursuing? Realists would have to answer in the affirmative. Since they believe in the existence of the theoretical entities posited by the best available theories, the result of this theoretical investigation could have a profound effect on their view of the world. But what would be the purpose of the exercise from an antirealist point of view? Fine suggests that the role of truth in realist explanations of scientific practices can always be played by empirical adequacy in an isomorphic antirealist account. But here the result of the scientific practice in question establishes the truth of a hypothesis for the realist (the hypothesis that a certain theoretical entity doesn't exist); yet there is no corresponding issue of empirical adequacy for the antirealist—by definition, the elimination that the theoretician is trying to effect will leave the empirical consequences of the theory unchanged. Thus, Fine's argument doesn't establish the universal availability of antirealist alternatives to realist accounts of scientific practices.

In cases where Fine's algorithm fails, however, the antirealist has an additional explanatory resource to call on. The algorithm fails for practices wherein the role played by truth doesn't involve any intermediary claims about empirical adequacy. But if empirical adequacy isn't at stake, the antirealist may be able to explain a scientific practice on *pragmatic* (as opposed to epistemic) grounds. For example, the switch from Ptolemaic to Copernican astronomy might be accounted for by antirealists by the greater ease of making calculations in the Copernican scheme. This appeal to computational facility won't work as an antirealist account of the scenario of the previous paragraph, for the elimination of an axiom can only make derivations more difficult. But there's no end to the pragmatic virtues that may motivate a piece of scientific work. Perhaps an antirealist will try to eliminate an empirically redundant postulate for the sake of maximizing beauty. After all, physicists have sometimes claimed that the aim of their theoretical work was to achieve beauty (e.g., Dirac 1963). If the appeal to beauty isn't a plausible pragmatic rationalization of some practice, then maybe an appeal to ugliness will do the job (our pragmatic purpose being to scandalize the bourgeoisie).

If it were possible to provide a pragmatic justification for every conceivable scientific practice, then we would arrive at Fine's conclusion via an entirely different route. However, the pragmatic gambit is not universally available. It works for all possible practices except those that have to do with the adoption of *beliefs*. For example, there can be a pragmatic justification for the practice of eliminating empirically redundant postulates. But the practice of *believing* in the theoretical entities posited by the minimal postulate set can't be justified pragmatically. A practice that involves believing something under certain conditions can be called an *epistemic* practice. By definition, there can be no pragmatic justification for epistemic practices.[1]

In the present context, we're particularly interested in the subset of epistemic practices that involve the adoption of *theoretical* beliefs—for example, the practice of believing in the theoretical entities posited by minimal postulate sets. As noted, these practices can't be given a pragmatic rationalization. Indeed, it's clear that there can't be *any* antirealist account of them, since to engage in them—that is, to be prepared to adopt theoretical beliefs—*is* to be a realist. No doubt these are the kinds of practices that Fine had in mind when he referred to "realist-laden" phenomena in his statement of Metatheorem 1. Beyond telling us that they're to be excluded from the range of the theorem, Fine has nothing to say about the realist-laden practices. Apparently, he regards it as obvious that their contemplation can carry no weight in the realism debate. But what is Putnam's claim in the conjunction argument, if not that the practice of theory conjunction is realist-laden? If we can specify a practice that is at once realist-laden and essential to the conduct of science, then the realist wins (assuming that science is rational). Strictly speaking, Metatheorem 1 is true as it stands: to every realist explanation of a *non-realist-laden* phenomenon, there corresponds an antirealist explanation of the same phenomenon. But because of its exclusion of realist-laden phenomena, Metatheorem 1 fails to underwrite the moral that Fine wishes to draw—that there's no point looking for phenomena that realists can account for but antirealists can't. For all that Fine tells us, it's possible that someone might discover a bona fide realist-laden practice. If there is such a practice, however, its realist-ladenness must not be at all obvious to the naked eye, for realists have searched far and wide for it and come up empty-handed.

Let's turn now to the second question of this section—whether there are practices that are compatible with antirealism but not with realism. The answer here is a mirror image of the answer to the first question: not counting "antirealist-laden" practices, a realist can do anything that an antirealist can do. The reason is that realists can, without logical impropriety, have any of the pragmatic interests that might activate an antirealist. They, too, can engage in scientific activities for the love of beauty, money, or predictive power. They remain realists nonetheless, as long as they're *additionally* willing to believe in some theories under some actualizable circumstances. This argument applies to every possible practice that an antirealist might engage in except for the antirealist-laden ones, such as the practice of *refusing to believe in theoretical entities under any circumstance*. The remarks made about realist-laden practices apply here with equal force, to wit: no one has ever succeeded in establishing the antirealist-ladenness of any practice that is essential to the conduct of science.

In sum, as far as is known at present, every scientific practice is compatible with both realism and antirealism. However, contra Fine, it has not been demonstrated that there can't be a practice that presupposes either doctrine.

3.3 Van Fraassen's Scientific Practice Argument

Granting that there's no inconsistency between any scientific practice and either realism or antirealism, it's still possible that one of these doctrines provides a *better* account of practice than the other. This is what van Fraassen and Fine claim for antirealism. According to van Fraassen, constructive empiricism accounts for any scientific practice that realism can account for, and it does so "without inflationary metaphysics" (1980, 73). I'm uncertain whether van Fraassen concedes that all scientific practices are compatible with realism. If his argument is otherwise sound, however, he can afford to concede this point. According to Fine, "no further work is done by ascending from [empirical adequacy] to the realist's 'truth'" (1986, 154). What is common to both van Fraassen's and Fine's remarks is the idea that if realism and antirealism can both account for the phenomena of scientific practice, then we should prefer the antirealist account on the grounds that it's a weaker theory. Constructive empiricism commits us to saying only that we believe in the empirical consequences of the theories we accept, whereas realism commits us to saying that we believe the entire theories. Both beliefs provide rational grounds for doing science the way we do it. But constructive empiricism does it via the imposition of fewer epistemic demands.

Now it can't be denied that belief in the empirical adequacy of T is logically weaker, hence more probably true, than belief in T. But this claim doesn't by itself justify the conclusion that we should believe only in empirical adequacy. After all, given any contingent hypothesis H, there are always weaker alternatives available, such as the disjunction of H and any other contingent hypothesis that is logically independent of H. If the existence of logically weaker alternatives blocked us from believing a hypothesis, then we could never believe anything except tautologies. On the other hand, it is appropriate to reject the less probable of two *incompatible* hypotheses: if $p(H)$ is greater than $p(-H)$, then we should definitely not believe $-H$. But isn't this admission enough to carry van Fraassen's argument to its conclusion? Isn't realism obviously incompatible with *anti*realism? And if that's so, then shouldn't the greater probability of antirealism lead us to reject realism? Perhaps the equivocation in this line of argument is too obvious to need spelling out. However, I believe that it accounts for much of the intuitive appeal of van Fraassen's argument. The problem is that the doctrine that is more probable than realism is not the same as the doctrine that is incompatible with realism. Let's define *realism* as the view that some theories (i.e., hypotheses referring to theoretical entities) are to be believed, and *antirealism* as the view that no theories are to be believed. Then realism and antirealism are indeed incompatible doctrines — it's logically impermissible to accept the truth of both realism and antirealism. But the van Fraassen argument fails with these definitions because antirealism is *not* logically weaker than realism. There's no way to derive the proposition that no theories are believable from the proposition that some theories are believable. What *is* logically weaker than realism is what might be called *arealism*, namely, the proposition that the empirical consequences of some theories are to be believed. There can be no doubt that arealism is weaker, hence more probable than, realism. The trouble is that arealism isn't incompatible with realism. On the contrary, every realist is also an arealist (which indicates that "arealism" is a bad name for this position), although some arealists may not be realists. In any case, the greater logical strength of realism is not, by itself, a reason to repudiate it. The van Fraassen argument, I think, plays on this equivocation. The term "constructive empiricism" is used to refer sometimes to the arealist view that we should believe in the empirical adequacy of theories, and sometimes to the antirealist view that we should believe *only* in the empirical adequacy of theories. In its former sense, constructive empiricism is seen to be a weaker hypothesis than realism. In its latter sense, constructive empiricism is an incompatible alternative to realism. QED.

I don't mean to suggest that van Fraassen actually makes this bad argument outright. In *The Scientific Image*, he claims that constructive empiricism accounts for any scientific practice that realism can account for and that, additionally, it is a weaker theory. He concludes from this that we ought to reject realism. Stated as such, the argument is a non sequitur: the higher probability of one of a pair of hypotheses is not a reason for rejecting the other. My discussion in the preceding paragraph aimed to show that the lacuna in the argument can't be filled in by an appeal to the incompatibility of realism and constructive empiricism. Van Fraassen didn't make this bad argument. But he did, in 1980, leave the hole in the argument unfilled. The hole remained vacant until 1985, when, in reply to his critics, van Fraassen gave the following reason for rejecting realism:

> If I believe the theory to be true and not just empirically adequate, my risk of being shown wrong is exactly the risk that the weaker, entailed belief will conflict with actual experience. Meanwhile, by avowing the stronger belief, I place myself in the position of being able to answer more questions, of having a richer, fuller picture of the world, a wealth of opinion so to say, that I can dole out to those who wonder. But, since the extra opinion is not additionally vulnerable, the risk is—in human terms—illusory, and *therefore so is the wealth*. It is but empty strutting and posturing, this display of courage not under fire and avowal of additional resources that cannot feel the pinch of misfortune any earlier. What can I do except express disdain for this appearance of greater courage in embracing additional beliefs which will *ex hypothesi* never brave a more severe test? (255)

If van Fraassen's disdain is elevated to the level of an epistemological principle, it might look something like this: when two hypotheses are *empirically equivalent* and one is logically weaker than the other, we should repudiate the stronger one. I look closer at this passage in chapter 7, where I try out a somewhat different reading. But this preliminary version will do for present purposes. In the present context, the hypotheses in question aren't first-order scientific theories—they're *metatheories* about the theoretical activities of scientists. Such metatheories may still have implications that are vulnerable to the dictates of experience. For instance, Putnam believed that the antirealist theory of scientific practice foundered on the empirical fact that scientists conjoin theories. What was argued in section 3.2, however, and what van Fraassen implicitly concedes, is that the realist and antirealist metatheories both have the same implications for scientific practice, or at least that there's as yet no reason to suppose otherwise. Thus, van Fraassen's 1985 principle fills the gap in his 1980 argument from scientific practice: if it's true that realism and weak, arealistic constructive empiricism both have the same consequences for scientific practice, then it follows by the 1985 principle that we should reject the realist account, that is, that we should become constructive empiricists in the strong, antirealist sense. Since van Fraassen hasn't offered us any other suggestions for how to plug the 1980 hole, I assume that his defense of the scientific practice argument must rely on the 1985 principle.

There are two problems with this reconstruction of van Fraassen's argument from scientific practice. The first is that there's no reason in the world why a realist should adopt the 1985 principle. Why should one be compelled to reject the stronger of two empirically equivalent theories? To be sure, we can't believe the stronger and reject the weaker. But why not believe both? The temptation to adopt the 1985 principle can only come from the belief that only the empirical consequences of a theory should count toward its beliefworthiness. But this is just what realists deny. Realists think that there are properties of theories other than their empirical consequences — properties such as simplicity or explanatory power — that should count as reasons for belief. The van Fraassen argument from scientific practice merely begs the question against them. There is, in fact, a more recent van Fraassen who seems to admit that he has no telling argument against the realists (van Fraassen 1989). These later views are discussed in chapter 12. In this chapter, I wish only to assess claims to the effect that there is an argument from scientific practice that favors either realism or antirealism. In 1980, van Fraassen claimed to have such an argument. I think he was mistaken. In view of his kindlier assessment of realism in 1989, it's likely that van Fraassen himself thinks that the 1980 argument was mistaken. But he hasn't stated in print what it was that made him give up his earlier and more negative appraisal of realism.

Now for the second problem with van Fraassen's argument. Let's suppose that realists could be persuaded to accept the 1985 principle. Then antirealism would follow as a consequence of the argument from scientific practice. But the argument would have the peculiar property that all references to scientific practice in it are superfluous. If it's granted that we should reject the stronger of two empirically equivalent theories, then the rejection of realism follows forthwith from a much more general argument that needn't refer to scientific practices at all. Indeed, it's the more general argument that van Fraassen defends in 1985. This argument runs as follows (recall that T* is the proposition that the empirical consequences of T are true):

(GA1) For any hypothesis T, T* is empirically equivalent to T.

(GA2) But T* is logically weaker than T.

(GA3) Therefore, we should believe only T*.

The missing assumption in this argument is supplied by the 1985 principle that we should repudiate the stronger of two empirically equivalent theories. This is the assumption that the realist will not accept. The argument from scientific practice can be represented as follows:

(SP1) Constructive empiricism (in its weak, arealistic sense) can account for any scientific practice that realism can account for (and, presumably, vice versa).

(SP2) But weak constructive empiricism is logically weaker than realism.

(SP3) Therefore we should believe *only* in weak constructive empiricism, which means that we should be constructive empiricists in the strong sense.

The missing premise in argument SP is the same one as in argument GA. Once again, the realist will not grant the truth of this additional premise. That was the first problem with van Fraassen's argument. The second problem is that even if the argument did work (i.e., if realists could be persuaded to accept the missing premise), it wouldn't accomplish anything useful. If it worked, then so would the general argument that makes no reference to scientific practice. The argument from scientific practice is just a *special case* of the more general argument, where T = realism, T* = weak constructive empiricism, and the empirical consequences of both theories are the practices of scientists. The fact that T and T* in this case are metatheories about theories, rather than first-level scientific theories, doesn't affect the structure of the argument. So the argument from scientific practice doesn't accomplish anything even if it's sound.

To summarize the results of the preceding sections: arguments from scientific practice have accomplished nothing for either side of the realism-antirealism debate. There's no reason to believe that there are any scientific practices that are incompatible with either realism or antirealism. There's no reason to believe that either side can provide a better account of actual scientific practice than the other. And if one side *could* be shown to provide a better account of scientific practice than the other, the same conclusion might very well be obtained by means of a general argument that renders any reference to scientific practice superfluous. For the most part, realists and antirealists seem to have agreed that a careful consideration of what scientists actually do will play an important role in the resolution of their debate. This belief is shared by all the philosophers mentioned in this chapter except Fine. Indeed, the idea that philosophical theses about science can only be settled by a careful study of the actual history of science has attained the status of a contemporary dogma. Its apotheosis occurs in the work of Ronald Giere (1985b). I think that the importance of history of science to philosophy of science has been greatly exaggerated. At any rate, its relevance to the realism debate has yet to be established.

3.4 A Clutch of Circularities

There's an instructive parallel between van Fraassen's argument for antirealism (SP) and the success of science argument for realism. I'll discuss the latter in its canonical "miracle argument" form. The miracle argument can be represented as follows:

(MR1) Realism explains the predictive success of science.

(MR2) Therefore, we should be realists.

This argument is vulnerable to exactly the same pair of criticisms that were leveled against van Fraassen's argument. First, both SP and MR beg the question. In fact, they both beg the same question, namely, whether nonempirical virtues like explanatoriness have epistemic import. Van Fraassen's scientific practice argument won't work unless we assume that nonempirical virtues *don't* have epistemic import, and Putnam's miracle argument won't work unless it's assumed that the nonempirical virtue of explanatoriness *does* have epistemic import (see section 2.4).

Second, it's an interesting fact about both SP and MR that even if these arguments worked, they would be superfluous. For SP to work means that some way has been found to get the realist to concede that the nonempirical virtues of a hypothesis can never provide us with a reason for believing it. But, as shown in section 3.3, if that concession were made, then there would be a more general argument, GA, that leads to the same conclusion without making any reference to scientific practice. The miracle argument yields to an entirely symmetric critique. For MR to work means that some way has been found to get *anti*realists to concede that explanatoriness *is* an epistemic virtue. But if that concession were made, then there would be a more general argument that leads to the same conclusion without making any reference to the predictive success of science as a whole. Let T be any theory (or metatheory) that has had an impressive record of predictive success. The general argument for realism would then look like this:

(GR1) The truth of T explains the truth of T^*.

(GR2) Therefore, we should believe T, which makes us realists.

The miracle argument is nothing more than a special case of the general argument where T is the metathesis of realism. So, questions have been begged and arguments superfluously multiplied on both sides in the realism-antirealism debate.

The question-begging and superfluous multiplication do not stop here. Fine (1986) offers the following critique of the miracle argument:

> In so far as realism might function in successful explanations of scientific practice, . . . success would give us grounds for believing in realism's central theoretical entities — correspondence, or real-World reference. Thus we treat "correspondence" analogously, say, to "electron", and count the explanatory success of theories that employ it as evidence for its "reality". But, of course, since this is precisely the pattern of inference whose validity antirealism directly challenges at the level of ordinary scientific practice, one could hardly hope to get away with using the same inference pattern at the meta-level. (161)

This passage contains my criticism of the miracle argument almost in its entirety. The charge of circularity is there, and the fact that the argument is a special case of a more general argument that renders it superfluous is strongly hinted at. So far, so good. But Fine endorses another argument to the effect that, problems of circularity and superfluity aside, antirealism offers a *better* account of the success of science than realism does. We've seen this argument already — it's a special application of his "proof" of Metatheorem 1, according to which there's a better antirealist account corresponding to *every* realist account. As applied to the miracle argument, the supposedly superior antirealist accounts runs as follows:

(MA1) Antirealism and realism both explain the predictive success of science.

(MA2) Antirealism is logically weaker than realism.

(MA3) Thus, antirealism is a better account of the success of science.

The basis for the first premise is the fact that the empirical adequacy of scientific theories already accounts for their predictive success. Note that Fine's has exactly the same structure as argument SP, van Fraassen's argument from scientific practice. Indeed, both Fine's MA and van Fraassen's SP are special cases of the more general argument GA for antirealism, also given earlier. Argument GA, in turn, is essentially Fine's argument for Metatheorem 1. Now, the point has already been made that SP and GA both beg the question against the realists. It therefore isn't surprising that Fine's argument MA, which is another special case of GA, perpetrates the same circularity. The missing assumption in MA is, once again, that only the empirical consequences of theories should count toward their beliefworthiness. In this case, the theories are realism and antirealism, and their empirical consequence is the predictive success of science. But this is to presuppose, once again, that the antirealist is right. Perhaps a brief catalog of circularities is in order: van Fraassen's argument from scientific practice begs the question in favor of antirealism, Putnam's miracle argument begs the question in favor of realism, and Fine's criticism of the miracle argument begs the question in favor of antirealism. It's almost enough to make one adopt the natural ontological attitude!

3.5 What's Wrong with the Natural Ontological Attitude?

Despite my repudiation of one of his arguments, I agree with Fine that there's no comfort for realists in the miracle argument, or in any other arguments that simply presuppose that some nonempirical virtue like explanation has epistemic significance. In the end, Fine also maintains that there is no case for antirealism. But there's a problem in reconciling this assessment with Metatheorem 1. How can one say that antirealism "has no argument" (1986, 168) if it's true that "to every good realist explanation there corresponds a better antirealist one" (154)? I suppose that the claim that antirealism has a better case is merely provisional, to be withdrawn when it's realized that the antirealist account simply begs a different question. Be that as it may, Fine must be faulted at least for not making the provisional nature of the conclusion, and its subsequent withdrawal, explicit. I also have a quibble about his claim that the antirealist has no argument. I prefer to say that antirealism has arguments — van Fraassen's argument from scientific practice is one — but that the arguments we've seen so far are circular. The difference between having a circular argument and having no argument is, of course, insignificant. But Fine's exposition misleadingly suggests that there's a lingering asymmetry between the realist's and the antirealist's case: the former is guilty of begging the question (1986, 161), while the latter has no argument (168). In fact, given what we've seen so far, they're in precisely the same dialectical boat.

    In the end, I agree with Fine that neither side in the realism debate has gained an advantage over the other. However, I'm not yet persuaded that the unrelieved exchange of question-begging arguments impels us to adopt the "natural ontological attitude" (NOA) that Fine recommends. To endorse NOA is to subscribe to no more than the "core position" on which both realists and antirealists agree, namely, that "the results of scientific investigations [are] . . . 'true', on a par with more homely truths" (1984, 96). There are three reasons why I resist becoming a NOAer. First, Fine's arguments against realism and antirealism are not so definitive as to undermine all hope of finding an advantage for one side or the other. As we've seen, his claim that to every good realist argument there corresponds a better antirealist argument is suspect, for his discussion doesn't rule out the possibility that there are essential practices that are realist-laden (or antirealist-laden, for that matter). Even if we're persuaded that there can't be a telling argument from scientific practice, it's still possible that some argument unrelated to scientific practice might do the job. Even if every extant argument were irreparably flawed (which I believe to be the case), it's possible that the next one to be devised will do the job. However, I admit that the prospects are grim.

My second and third reservations about NOA are, I think, considerably weightier. The second is that the term "natural ontological attitude" is susceptible to the same seductive equivocation as "antirealism" and "constructive empiricism". Fine maintains that NOA is truly deflationary — that is, that it really is weaker than either realism or antirealism. This claim is akin to van Fraassen's assertion, which Fine repudiates, that constructive empiricism is deflationary vis-à-vis realism, and it involves the same slide from a weaker to a stronger sense of the operative term. *Weak* NOA is believing in the core position on which realists and antirealists agree; *strong* NOA is believing in *no more than* the core position on which realists and antirealists agree. It's only weak NOA that is deflationary regarding realism and antirealism. But weak NOA isn't incompatible with realism or antirealism. On the contrary, both realists and antirealists possess the weak natural ontological attitude. Since Fine is exhorting realists and antirealists to change their ways, I can only assume that he espouses strong NOA, as well as weak. But *this* doctrine is not deflationary regarding realism or antirealism. So, what is the argument for strong NOA? What's the basis of its purported superiority to realism and antirealism? Of all the different points at which our circle of beliefs may stop, why should we designate the boundary of the "core position" as special? Why should we not step beyond the core? Compare Fine on constructive empiricism:

> Constructive empiricism wants us to restrict belief to empirical adequacy because (we are told) the only way experience could count against a theory is (logically speaking) by first counting against its empirical adequacy. . . . But why this Popperian twist to the point? Why should the fact that empirical adequacy is first in the line of vulnerability to experience issue in a blank policy of restricting belief exactly there? (1986, 168)

According to Fine, van Fraassen does not answer these crucial questions: "constructive empiricism has no argument" (168). What is Fine's answer to the corresponding question about NOA?

His answer is that the core position provides everything that science "needs" to conduct its business. Anything more is a superfluous "extrascientific commitment" to a "philosophical school" (1986, 171). This reply suffers from the same arbitrariness as van Fraassen's position. Why should we restrict our beliefs to those that belong to the intersection of the proscience attitudes of realists or antirealists? Why not let some antiscience points of view into the circle — say, the views of biblical fundamentalists? The "natural ontological attitude" might just as well be defined as the core position common to all those who have any epistemological stance whatever. Alternatively, why not make the club more exclusive? Maybe we should restrict our belief to the core position shared by all *physical* scientists. I don't deny that we may rationally *choose* to restrict our circle to those who share a pro attitude toward science. But this isn't to say that such a decision is more rational, or more optimal, than its rivals.

My third and final criticism of NOA is that it suffers from a critical vagueness stemming from our inability to say where science leaves off and philosophy begins. The distinction between science and philosophy seems crucial for Fine's purposes, since NOA eschews both realist and antirealist stances on the grounds that they are "extrascientific commitments". Let us remember that the old "demarcation problem" has never been solved. Like most philosophical dilemmas, it merely faded away. For my part, I don't see why the hypothesis that there are unobservable entities in the universe should be regarded as qualitatively different from run-of-the-mill scientific hypotheses. What sets it apart? The fact that scientists disagree about realism doesn't set it apart, since scientists routinely disagree about run-of-the-mill issues. Nor does our inability to resolve the problem by appealing to data—for neither can the problem of choosing between any two curves that go through the same set of data points—and what could be more run-of-the-mill than that? So what makes the realism issue so different as to place it beyond the pale of some ideal of reasoned discourse?

In sum, I don't (yet) have the natural ontological attitude. I share with that attitude the belief that there are (as yet) no good reasons for endorsing either realism or antirealism. But, unlike NOAers, I'm not committed to the view that the issue is irresolvable. In fact, I'm not (yet) committed to anything except the repudiation of a lot of ineffectual arguments. This is not an attitude rich enough to warrant baptizing with a special name (I develop an attitude with a name before ending the book, however). But that's what a really deflationary stance is like.

Realism and
Theoretical Unification

Michael Friedman (1983) has presented an ingenious analysis that at once pro-vides a new argument for scientific realism and explains why theoretical uni-fication is a valued scientific accomplishment. His argument for realism, in broad outline, runs as follows:

(F1) Theoretical unification results in better confirmed accounts of the world than we would otherwise be able to obtain (this is why unification is valued).

(F2) Only scientific realists have rational grounds for unifying theories.

(F3) Therefore, we should be realists.

In section 4.1, I claim that this argument is defective on two counts: the argument is circular, and F2, the second premise, is false. However, subsequent sections show that F1 is true and that its truth has an interest and significance that is independent of the role it plays in the argument for realism. Friedman's argument for F1 makes reference to a mode of theoretical reasoning that has been largely neglected by phi-losophers of science. Friedman maintains that this mode of reasoning accounts for the epistemic benefit reaped by a successful theoretical unification. I agree that Friedman has succeeded in capturing one reason why some instances of unification constitute scientific progress (section 4.2). However, in section 4.3 I claim that there are epistemic benefits to some types of unification that require a different account. I then show that Friedman's account and mine are two instances of a broader category of theoretical strategies, none of which has yet received its philosophical due (sec-tion 4.4). Finally, I close with a comparison between my account of the varieties of theoretical strategies and Laudan's (1977) seminal analysis of "conceptual problems" (section 4.5). Some aspects of this protracted meditation on F1 figure in the analyses of following chapters. In particular, the mode of reasoning used by Friedman to establish F1 reappears in a refutation by Laudan and Leplin of a famous argument for antirealism (see section 6.2). However, a lot of the material in this chapter has no immediate bearing on the realism issue. It's here because it emerges directly from the contemplation of F1.

A caveat: some readers of earlier versions of this chapter objected to my presentation of Friedman's argument in terms of theoretical probabilities. They claimed that Friedman was talking about theories' *degrees of confirmation*, and that, like Glymour (1980), he didn't regard the confirmation function as a probability function. But there's at least one place in Friedman's discussion in which a theory's "degree of confirmation" is explicitly compared with the "prior probability" of its empirical consequences (1983, 244). More important, my arguments against Friedman don't make use of any properties of probabilities that Friedman doesn't explicitly ascribe to "degrees of confirmation." Specifically, the property that is crucial to the argument is that no claim can have a smaller probability/degree of confirmation than a hypothesis from which it can logically be derived. In brief, the arguments against Friedman wouldn't be affected if "degree of confirmation" were substituted for "probability" throughout.

4.1 Friedman's Argument for Realism

Friedman's argument gives a novel twist to Putnam's (1975b) conjunction argument for scientific realism. The original conjunction argument is described in chapter 3, but it won't hurt to run through it again. Suppose that T_1 and T_2 are accepted theories, and that the empirical claim E is a consequence of the conjunction T_1 & T_2. In that case, scientists will, of course, believe that E is true. But this inference presupposes that realism is true. For, if antirealism is true, the acceptance of T_1 and T_2 would license no more than a belief in the empirical adequacy of T_1 and T_2, and from the fact that the empirical consequences of T_1 and T_2 are true, one cannot conclude that the empirical consequences of T_1 & T_2 are also true. Therefore, scientific practice presupposes realism.

There are a number of ways in which this argument may be criticized. Van Fraassen's (1980) response has been to deny that scientists ever do conjoin theories in this straightforward manner. Should this argument fail, the antirealist may still be able to avoid Putnam's conclusion by biting the bullet and asserting that the scientific practice of theory conjunction is irrational. This isn't a live option for philosophers who take the practices of science to be *constitutive* of rationality. But it would be nice for realists if they didn't have to rely on this arguable assumption. Friedman's argument circumvents both potential sources of difficulty for Putnam's argument. Friedman contends that there are normative reasons why scientists *should* believe in the consequences of conjoint theories. He also claims that scientists do routinely conjoin theories. But the latter claim doesn't play an essential role in his argument, as it does in Putnam's. If it turned out that most scientists avoided theory conjunction because of antirealist scruples, Friedman's argument would lead to the conclusion that they should change their ways.

The argument runs as follows. For any theory T, let T^* be the hypothesis that T's empirical consequences are true. Now consider the situation of a scientific realist who is willing to believe T, and an antirealist who is only willing to believe T^*. Both of them may submit T to empirical testing—that is, derive an empirical consequence E of T and ascertain whether E is true. If E does turn out to be true, both of them may increase p(T*), the probability that T is empirically adequate. The realist will also wish to elevate p(T), the probability that the theory itself is true. The antirealist systematically refuses to take this step. Whom should we emulate?

Friedman claims that realism gives one an epistemic advantage that the antirealist is forced to acknowledge. As long as we restrict our attempts at confirming T to empirical consequences of T in isolation, there's no reason to expect a divergence of opinion about p(T*) between realists and antirealists. The fact that confirmation also leads the realist to augment p(T) is *not* a consequence of realism that the antirealist is forced to recognize as advantageous. But the realist can rationally *conjoin* T with other theories to derive additional empirical consequences, the truth of which may provide additional confirmation for T, causing an even greater elevation of p(T):

> A theoretical structure that plays an explanatory role in many diverse areas picks up confirmation from all these areas. The hypotheses that collectively describe the molecular model of a gas of course receive confirmation via their explanation of the behavior of gases, but they also receive confirmation from all the other areas in which they are applied: from chemical phenomena, thermal and electrical phenomena, and so on. (1983, 243)

The antirealist will not, of course, be swayed by this consequence of realism. But if correct predictions from T conjoined with various other theories continue to pile up, p(T) will become so large—*for the realist*—that it will exceed p(T*):

> By contrast, the purely phenomenological description of a gas . . . receives confirmation from one area only: from the behavior of gases themselves. Hence, the theoretical description, in virtue of its far greater unifying power, is actually capable of acquiring more confirmation than is the phenomenological description. (243)

Friedman continues:

> This may seem paradoxical. Since the phenomenological description of a gas is a consequence of its theoretical description, the degree of confirmation of the former must be at least as great as the degree of confirmation of the latter. My claim, however, is not that the phenomenological description is less well-confirmed than the theoretical description after the former is derived from the latter—this, of course, is impossible. Rather, the phenomenological description is less well-confirmed than it would be if it were *not* derived from the theoretical description but instead taken as primitive. The phenomenological description is better confirmed in the context of a total theory that includes the theoretical description than it is in the context of a total theory that excludes that description. This is because the theoretical description receives confirmation from indirect evidence—from chemical phenomena, thermal and electrical phenomena, and the like—which it then "transfers" to the phenomenological description. (243–244)

The concept of "transferring" confirmation from the theoretical to the phenomenological description is explicated as follows:

> I am not claiming that every increase in the degree of confirmation of the theoretical description leads to an increase in the degree of confirmation of the phenomenological description. I do not endorse the dubious "consequence condition".

> . . . I am only pointing out that if the degree of confirmation of the theoretical description is *sufficiently* increased (for example, if it actually exceeds the prior probability of the phenomenological description), the degree of confirmation of the phenomenological description will be correspondingly increased as well. (Here I am indebted to Adolph Grünbaum.) (244)

In other words, $p(T)$ may get so high, as a result of the confirmations that T picks up from various successful conjunctive predictions, that it becomes greater than $p(T^*)$. But T^* is a logical consequence of T; therefore it's incoherent to suppose that $p(T) > p(T^*)$. The requirement of probabilistic coherence thus forces us to elevate $p(T^*)$.

But, Friedman argues, this consequence of realism is one that *should* count for something among antirealists. Apparently, the realist ends up with a better confirmed view of *observable phenomena* than the antirealist does. Realists therefore have firmer knowledge of matters concerning which antirealists *concede* we are trying to obtain knowledge. Friedman takes this to be a good reason for accepting realism, for "if we give up this practice [of conjoining theories], we give up an important source of confirmation" (246).

This argument has been criticized by Morrison (1990) on the grounds that it begs the question. Friedman claims that the benefit of conjunction is an argument in favor of realism. But, according to Morrison, you have to *be* a realist to begin with to reap this benefit:

> Since both [Friedman] and Putnam claim that our theories evolve by conjunction it appears that we must have some *prior* belief in the truth of our hypotheses in order to achieve the desired outcome. So, although Friedman cites unification as a justification for the literal interpretation of theoretical structure, it is interesting to note that on his account we cannot achieve a unification unless we *first* adopt a reductivist approach that construes the theoretical structure as literally true. In other words, in order to have a unified theoretical structure we must be able to conjoin our theories, which in turn requires the belief that they are true; but this was the same belief for which unification was thought to provide a justifying condition. Hence it appears as though we cannot simply limit belief to the unifying part of the theory, we need a stronger form of realism to motivate this model of unification. (310)

The problem, according to Morrison, arises because we have to decide *which* theories, or which portions of a given theory, to be realist about. Friedman's only reply to this question is that we should believe in those theories (or portions of theories) that can successfully be conjoined with other theories. But the theories that we take to be conjoinable are precisely the ones that we are realists about. Thus, Friedman's thesis reduces to the advice that we should be realists about the theories that we are realists about.

Stated in this way, I think that Friedman has an adequate reply to the charge of circularity. Let's say that antirealism is the doctrine that one should always ascribe either zero probabilities or totally vague probabilities to theories that make non-observational claims. In contrast, realism is the willingness to ascribe discrete non-zero probabilities to some theories of this type. Friedman's argument for realism may be construed as the claim that realists will in *some* instances end up with a firmer knowledge of observational truth than antirealists will be able to acquire. To be sure, the realist cannot know ahead of time which theories will prove to be conjoinable with other theories. If the consequences of T in conjunction with other theories turn out to be false, then belief in T (or in the other theories) will have to be given up. But that leaves the realist in no worse a position, vis-à-vis T, than the antirealist. On the other hand, if T does prove to be conjoinable with other theories, then the realist's knowledge of T* will be more firmly established than the antirealist's. Heads, the realist wins; tails, it's a tie. Therefore, it's wise to be a realist.

There's another way to prosecute the charge of circularity, however. According to Friedman, the advantage of realism is that realists end up with better established beliefs about the phenomenal world. But it's question-begging to suppose that this state of affairs is an *advantage* unless we have independent grounds for supposing that the beliefs in question *should* be given greater credence. Consider the philosophical stance of *gullibilism*, which has the same putative advantage as realism. Gullibilism is the view that one should (1) ascribe extremely high prior probabilities to every theory, (2) boost these probabilities generously when theories are confirmed, and (3) diminish them minimally when they're disconfirmed. To avoid incoherence, gullibilists will have to adhere to a number of additional stipulations. One problem with their methodology is that it's impossible to give this kid-gloves treatment to both a theory and its negation. The type of gullibilist that I have in mind solves this problem by giving preferential treatment to the first hypothesis in the field: if T is proposed before −T, then T is given a high prior probability, and if −T is proposed before T, then it's −T that gets the high probability. Other conceptual problems with gullibilism can be resolved by additional ad hoc devices. Despite the adhockery, gullibilists will enjoy precisely the same "advantage" over garden-variety realists that the latter are supposed to enjoy over antirealists: they will have much firmer opinions about matters concerning which both they and their opponents are trying to gain knowledge. The problem, of course, is that there's no reason to believe that these firmer opinions are warranted.

The same can be said by antirealists in response to Friedman's argument. Given the realist's ascription of a nonzero prior probability to T, it's possible that $p(T)$ becomes so great that we're forced to augment $p(T^*)$. Given the antirealist's ascription of a zero probability or a totally vague probability to T, $p(T)$ never gets anywhere, no matter how many successful conjunctions it partakes in; so, antirealists are never forced to augment $p(T^*)$ to make it keep pace with $p(T)$. But the realist's firmer opinions about the phenomenal world do not by themselves establish that the realist's conclusions are epistemically superior to the antirealist's: being more confident of one's views is not ipso facto an epistemic virtue. To count the greater confidence in T* as an advantage, one has to assume that the realist's prior probabilities are a better starting point than the antirealist's. But this is to beg the question.

Now let's put the problem of circularity aside and suppose that theory conjunction *is* epistemically advantageous in the way that Friedman suggests. Even so, the existence of this epistemic advantage doesn't provide us with a reason for realism, for the advantage can as well be reaped by some antirealists — namely, those dubbed "conjunctive empiricists" in chapter 3. Conjunctive empiricism is the willingness to believe in T#, where T# is the hypothesis that the empirical consequences of T, in conjunction with whatever auxiliary theories are accepted, are true. Conjunctive empiricism is stronger than constructive empiricism, but it's still a species of antirealism. Nevertheless, conjunctive empiricists have just as good a reason to conjoin theories as do realists: $(T_1 \& T_2)^*$ doesn't follow from $(T_1)^*$ and $(T_2)^*$, but $(T_1 \& T_2)\#$ is a logical consequence of $(T_1)\#$ and $(T_2)\#$. Therefore, even if we grant that the practice of conjunction is epistemically advantageous, this advantage is not a reason for realism.

It might be objected that the supposition that T# could be true without T also being true is too fantastic to be believed. I note in chapter 3 that this objection does nothing more than reiterate the discredited miracle argument. Appeals to the miracle argument would be especially inappropriate in the present context—for it isn't open to Friedman to dismiss conjunctive empiricism on the grounds of the miracle argument, and at the same time to suppose that a new argument for realism advances our understanding of the issues. If we had the miracle argument, we wouldn't *need* Friedman's argument.

4.2 Friedman's Account of Unification

To recapitulate: Friedman maintains that theoretical unification results in better confirmed phenomenological laws, that only realists have rational grounds for unifying theories, and that, as a consequence, we should become realists so that we can avail ourselves of this benefit. In section 4.1, I claimed that Friedman's argument for realism is question-begging. If the premises of the argument are correct, then realists and antirealists will have different opinions concerning the degree to which their phenomenological laws are confirmed. But to suppose that the realists' opinion on this matter is preferable to the antirealists' is to prejudge the issue between them. I also argued that even if the argument were valid, the second premise is in any case false, for antirealists may also have epistemic reasons for unification.

It's worth noting that there is nothing in the foregoing critique that calls to question the first of Friedman's premises—that for those scientists whose probability functions permit it, a successful unification may increase the confirmation level of the unified hypotheses. For realists, the confirmation of $T_1 \& T_2$ can result in a boost, not only to $p(T_1^*)$ and $p(T_2^*)$, but also to $p(T_1)$ and $p(T_2)$. For conjunctive empiricists, the confirmation of $(T_1 \& T_2)\#$ can boost $p(T_1\#)$ and $p(T_2\#)$. There may be types of antirealism that are so skeptical that the confirmational mechanism described by Friedman has no sphere of application. But it's incumbent on *everyone* to recognize the validity of what may be called the *Friedman maneuver*: if we know that a hypothesis H_1 entails another hypothesis H_2, and *if* H_1 receives so much confirmation that its probability level is boosted beyond that which is ascribed to H_2, then the requirement of probabilistic coherence dictates that we must elevate the probability of H_2. The only question is which hypotheses this can happen to. If one ascribes a prior probability of zero, or a totally vague probability, to certain types of hypotheses (e.g., those referring to theoretical entities), then no amount of confirmation will ever elevate their probabilities, and the antecedent condition for applying the Friedman maneuver will never be satisfied. This isn't a criticism of the rationality of the maneuver, however. Indeed, the validity of the maneuver can hardly be gainsaid. It follows directly from the laws of probability theory. As Friedman points out, the rationality of his maneuver explains why successful theoretical unifications are considered to be scientific advances. When new empirical consequences of T_1 & T_2 are confirmed, we become more confident of the truth T_1^* and T_2^* (and, if we are realists, of T_1 and T_2), even though no new consequences of T_1 or of T_2 have been confirmed. A successful unification gives us rational grounds for increasing our confidence in the hypotheses that are unified. Unification is therefore a method for putting hypotheses on a firmer epistemic footing other than by directly testing their empirical consequences.

The foregoing is actually Friedman's second attempt to explain what unification is for. According to his first account, the unification of two theories T_1 and T_2 by a single theory T_3 is to be valued because it decreases the total amount of mystery in the world (Friedman 1974). Prior to the construction of T_3, we had two irreducible truths about the world, T_1 and T_2, which had to be accepted as "brute facts". After T_3, we have only a single brute fact. Therefore, our total understanding of the world is greater than it was before. The main problem in establishing such a view, of course, is to specify the operative principle of individuation that enables us to count up brute facts. Friedman makes a valiant attempt at it, but I think that Salmon's (1989) critique establishes that his approach will not work. In any case, there's no mention of mystery or understanding in Friedman's second discussion of unification. On this second account, what makes unification desirable is its connection to *confirmation* rather than to understanding.

There's a palpable change of topics between Friedman's first and second discussions of these issues that points to an ambiguity in the term "unification". The kind of unification dealt with in the earlier article was the process whereby a new theory T_3 is constructed that (speaking loosely) entails two older theories T_1 and T_2. This is the sense in which Maxwell's electromagnetic theory unified the previously disparate theories of electricity and magnetism, or the electroweak theory unifies the electromagnetic theory and the theory of weak nuclear forces. The mere conjunction of two theories, T_1 & T_2, would *not* count as a unification of T_1 and T_2 in the 1974 article. Indeed, Friedman's earlier analysis couldn't even get off the ground if T_1 & T_2 was allowed to count as a unification of T_1 and T_2. Let's refer to the former type of unification as *grand* unification, as opposed to unification by mere conjunction. The confirmatory mechanism described in the later discussion — the Friedman maneuver — applies equally to *both* sorts of unification. The maneuver was described above in terms of conjunction. It hardly needs to be changed at all to make it applicable to grand unification: if we devise a brand-new theory T_3 that implies T_1 and T_2 (hence also T_1^* and T_2^*), then T_3 may receive so much additional confirmation that $p(T_3)$ comes to exceed $p(T_1^*)$, whereupon $p(T_1^*)$ must be increased to restore probabilistic coherence.

I accept the claim that this confirmatory mechanism accounts for the desirability of both types of unification. What it does *not* account for, however, is Friedman's 1974 intuition that grand unification provides epistemic advantages *over and above* those of mere conjunction. It seems compelling that such an additional advantage exists. If it didn't, then why would we even try to perform grand unifications, rather than simply assert the conjunction of all theories we believe in? I will offer a new account of the special advantage of grand unification in the next section. But first it's necessary to consider the claim, made by both Morrison and van Fraassen, that no such thing as grand unification ever occurs. Indeed, van Fraassen and Morrison maintain that neither grand unification nor unification by conjunction ever occurs, for the target theories always undergo modification prior to being unified. As noted in chapter 3, Trout's (1992) discussion of the mercenary use of theories as auxiliaries establishes that van Fraassen and Morrison are wrong about conjunctive unification. But what about grand unification? One problem with van Fraassen's and Morrison's thesis is that it seems to lead to the conclusion that there is no principled distinction to be made between grand unifying theories and run-of-the-mill new theories. If that were so, it would be difficult to understand why scientists have historically gotten excited about grand unifications. There is a way to reconcile van Fraassen's and Morrison's claim with the intuition that grand unification is special. Even if it's true that new unifying theories never preserve the content of their predecessors in toto, they must preserve *something* of the original theories, or else we wouldn't be able to recognize them as unifying theories. Without taking any position as to exactly what is preserved in a unification of T_1 and T_2 by T_3, let's refer to the preserved *kernels* of T_1 and T_2 as $T_3(T_1)$ and $T_3(T_2)$. $T_3(T_1)$ and $T_3(T_2)$ have at least the following properties: T_1 entails $T_3(T_1)$, T_2 entails $T_3(T_2)$, and T_3 entails both $T_3(T_1)$ and $T_3(T_2)$. The fact that T_1 and T_2 are modified before being unified certainly does not establish that $T_3(T_1)$ and $T_3(T_2)$ are not significant hypotheses. Indeed, most new theories will have *some* elements in common with their predecessors. On this view, the difference between grand unifications and run-of-the-mill new theories is a matter of degree: the more of T_1 and T_2 preserved in $T_3(T_1)$ and $T_3(T_2)$, the more completely T_3 unifies T_1 and T_2. The intuition that grand unification confers epistemic advantages over and above those provided by the Friedman maneuver is thus immune to van Fraassen's and Morrison's charge of vacuity. We just have to take account of the fact that grand unification comes in degrees.

But what *is* the advantage of grand unification over mere conjunction? One difference between the two is that grand unifying theories are generally able to correct inadequacies in the original theories. To use Friedman's favorite example, the kinetic-molecular theory of gases not only yields the Boyle-Charles law for gases under suitable idealizing assumptions; it also gives the more accurate van der Waals law when the idealizations are relaxed. This is certainly a mark of theoretical improvement. But there's nothing about this characteristic that depends on the unificatory character of the new theory. Furthermore, suppose that we find a single grand unifying theory T_u that has exactly the same empirical consequences as the conjunction of a long series of theories, $T_1, T_2, T_3, \ldots, T_n$, which were hitherto thought to be unconnected. Perhaps this has never happened in real science. But if it did, wouldn't it still be the sort of case in which we suppose that grand unification has a special epistemic advantage? That is to say, wouldn't we think that the construction of T_u is a contribution to theoretical progress? Certainly this is what Friedman claims in his earlier discussion of unification (1974, 14). In fact, the intuition that T_u represents an epistemic improvement over $T_1 \& T_2 \& \ldots \& T_n$ is precisely what his earlier discussion attempts to ground. This early attempt fails. The second discussion of unification in 1983 succeeds in identifying an epistemic advantage that applies equally to grand unification and conjunctive unification. But for that very reason, it also fails to identify any advantage pertaining to Tu but not to T1 & T2 & . . . & T*n*.

4.3 A Bayesian Account of Unification

An adequate account of the advantage of grand unification can be given within a Bayesian framework. For ease of presentation, let's consider only the dyadic case where T1 and T2 are unified by an entirely new theory T3. Let's also assume that the unification is perfect, that is, that T3 entails T1 and T2. A crucial feature of the Bayesian approach is that there are no objective constraints beyond coherence on the prior probability we assign to a new hypothesis. Suppose, then, that T3 initially strikes us as so plausible that we wish to assign a very high prior probability to it. Perhaps we think T3 is extremely simple and we have an epistemic predilection for simplicity. At the moment when we assign a prior probability to T3, we may not have in mind the fact that T3 unifies T1 and T2. In fact, this logical relationship between T3, T1, and T2 may not have been discovered yet, just as the relationship between the molecular theory and Brownian motion was not known for some time. Suppose, then, that the prior probability we ascribe to T3 is actually *greater* than the current probability of T1 & T2. Then, when we're apprised of the relationship between T3, T1, and T2, we realize that our probability function has become incoherent: since T3 entails T1 & T2, its probability can't be greater than p(T1 & T2). The requirement of probabilistic coherence dictates either that we diminish p(T3) or that we elevate p(T1 & T2). I have elsewhere discussed the unsolved problem of how to choose among the infinitely many ways that are always available for rectifying incoherence (Kukla 1995). In the present context, it's enough to say that there surely are *some* scenarios in which the indicated path is to elevate p(T1 & T2). These are the scenarios in which grand unification yields a new epistemic advantage. The assumption that the initial incoherence was due to our not being *cognizant* of the fact that T3 entails T1 & T2 was made only for didactic purposes. Even if we're aware that T3 implies T1 & T2 at the very moment that we assign a prior to T3, we may reason as follows: my present probability for T1 & T2 is p(T1 & T2), but the new theory T3 is so beautiful and so simple that I feel impelled to assign a probability p(T3) to it that is much higher than p(T1 & T2); therefore, to maintain coherence, I must elevate p(T1 & T2). Let's call this theoretical strategy the *Bayesian* maneuver.

A comparison between the Bayesian maneuver and run-of-the-mill theoretical improvement is instructive. Theoretical improvement in general involves the introduction of a new theory that has either a broader set of empirical consequences or a greater probability than its predecessors. The Bayesian maneuver always produces a theoretical improvement, since the prior probability of the new, unifying theory is always greater than the probability of the old theories that are unified. But the special epistemic benefit of the Bayesian maneuver is an additional bonus that is not available to any and every theoretical improvement. Theoretical improvements often have the result that the new theory *supplants* the old ones. When the new theory effects a grand unification of the old theories, however, the result is that our confidence in the old theories is increased. The presentation of a highly plausible theory

T3 that entails T1 & T2 is an argument for the acceptance of T1 & T2. Grand unification can be regarded as a method of increasing the probability of a theory without
doing any new empirical research.

To be sure, the foregoing analysis has been conducted under the idealizing assumption that the unifying theory T3 strictly entails the unified theories T1 and T2.
Let us now suppose that T3 entails only the kernels T3(T1) and T3(T2) of T1 and T2.
This still leaves scope for the operation of the Bayesian maneuver: if the prior probability of T3 is greater than the probabilities of the kernels, then the probabilities of
the kernels must go up. Partial unification still results in some aspect of our old knowledge being placed on a firmer footing. It is true that most—perhaps all—theoretical
improvements are partial unifications, in the sense that they preserve *something* of
their predecessors' claims. But even if we believe that all theoretical improvements
are partial unifications, we can still conceptually separate the effect due to the Bayesian maneuver from other epistemic benefits.

Now let's compare the Bayesian maneuver to Friedman's maneuver. In both cases,
unification results in an increase in the probabilities of the hypotheses that are unified.
In both cases, this elevation is due to the fact that the probability of the unifying theory
becomes greater than that of the unified theories. In the case of Friedman's maneuver,
however, this excess of probability is produced by the extra confirmation obtained by
the unified theory. In the case of the Bayesian maneuver, the excess of probability is
due to the fact that the new theory is prima facie more plausible than we had thought
the old theories to be. Friedman's maneuver applies equally to both grand unifying
theories and straightforward conjunctions; hence, it can't account for the special
epistemic advantage that we suppose grand unification to have. The Bayesian maneuver, however, applies only to grand unifications, for it's senseless to suppose that one
can assign a greater probability to T3 than to T1 & T2 when T3 just *is* the conjunction
T1 & T2[1]. Therefore, the Bayesian maneuver supplies us with the desideratum that
Friedman's maneuver fails to provide, namely, an epistemic advantage to grand unification that is not shared by conjunctive unification.

Finally, let's compare the Bayesian maneuver with Friedman's 1974 analysis of
grand unification. According to the latter, the benefit of grand unification is that it
diminishes the number of propositions that we have to accept as brute facts. Essentially, this is a species of the more general view that grand unification effects a simplification in our worldview. The persistent problem with appeals to simplicity in
every context has been that they employ criteria of simplicity that turn out to be
dependent on a choice of primitive terms, and that it proves to be impossible to give
a principled reason for choosing one set of primitives over another. Salmon's (1989)
critique shows that Friedman's method of enumerating brute facts falls prey to the
traditional difficulties. Now, the present account of grand unification *allows* that
judgments of simplicity may be a major determinant—perhaps even the only determinant—of the epistemic merit of the unifying theory. It's even possible that prior
probabilities are assigned to unifying theories on the basis of how much they reduce
the number of brute facts we have to assume about the world (with respect to some
canonical set of primitives). But there's nothing in the Bayesian maneuver that requires us to assert that the operative judgments of simplicity, or the choice of primitives, have objective validity. The objectivity of the procedure is rooted in the much

firmer ground of probabilistic coherence. To be sure, two scientists, S_1 and S_2, may disagree over whether the prior probability of a given unifying theory is in fact great enough to force an upward adjustment in the probabilities of the theories that are unified. But *if* they both find a T_3 such that T_3 entails T_1 & T_2 and $p(T_3) > p(T_1$ & $T_2)$, *then* they will both agree that $p(T_1$ & $T_2)$ needs to be increased. And this is why both S_1 and S_2 agree to the general principle that the construction of a *good* unifying theory, that is, one that most investigators will initially regard as highly plausible, brings a distinctive epistemic advantage in its train.

Does the Bayesian maneuver explain everything about grand unification? An affirmative answer would imply that every grand unified theory must be considered initially more plausible than the theories it unifies. Certainly this is true *some* of the time. I. G. Good (1968) gives an example. Prior to Newton's synthesis, the laws of planetary motion were considered to be extremely implausible. But Newton's law of gravitation, which entailed Kepler's laws, was immediately regarded as extremely plausible. According to Good, the construction of the law of gravity had the effect of removing the implausibility attached to the laws of planetary motion. This is a perfect example of the Bayesian maneuver. On the other hand, it doesn't seem reasonable to suppose that grand unifying theories are *always* considered initially more plausible than their predecessors. These cases, however, might be covered by the Friedman maneuver: if the grand unifying theories receive more confirmation than the theories they unify, then they eventually may lead to an increase in the probability of the latter, even if they began life as implausible hypotheses. But this is a potential advantage shared by conjunctive unifications. The question is whether there are any other epistemic advantages that pertain exclusively to grand unifications. If there are, then it should be possible to have a grand unifying theory that (1) is initially less plausible than any of the theories that it unifies (so the Bayesian maneuver is inapplicable), (2) has no empirical consequences beyond those of the conjunction of the theories it unifies (so the Friedman maneuver is inapplicable), yet (3) is considered nevertheless to be an epistemic advance. My intuition tells me that such a grand unification would be a waste of time. If this intuition is right, then the only reason for trying to effect grand unifications is so that we can perform the Bayesian maneuver.

4.4 Evolving Probability and Scientific Progress

The conceptual similarity between the Friedman maneuver and the Bayesian maneuver is striking. In fact, both maneuvers belong to a broader category of theoretical strategies that have received very little philosophical press. A general description of the strategies is quickly given: an incoherence in our probability function is discovered, or produced, or threatened, which in turn forces us to adjust our beliefs in a way that restores or maintains coherence. As far as I know, it was Good (1968) who first noted that this kind of probabilistic adjustment is a routine event in the history of science. Good called the phenomenon *evolving probability*. Friedman's and the Bayesian maneuver are both manifestations of evolving probability. In the case of Friedman's maneuver, we have a unifying theory that receives extra confirmation not available to the theories it unifies. If this extra confirmation is so great that the probability of the unifying theory threatens to exceed the probability of the theories that are unified, we must increase the latter to avoid incoherence. In the case of the Bayesian maneuver, we have a unifying theory that is simply prima facie more plausible than the theories it unifies. Once again, the result is that we must increase the probabilities of the theories unified in order to avoid incoherence.

The circumstances that are suitable for the application of the Friedman and Bayesian maneuvers are interestingly different. In the Bayesian maneuver, what produces the incoherence is our very first assessment of the new theory's plausibility. In fact, theoreticians who employ the Bayesian maneuver may have that maneuver in mind as they are working on a unifying theory: they are *looking* for a unifying theory that will exceed its predecessors in beauty and simplicity, hence, for most scientists, in plausibility. The Bayesian maneuver may be described as the *invention* of a theory that entails its predecessors but is more plausible than its predecessors. In the case of the Friedman maneuver, however, the unifying theory has to receive multiple confirmations before the maneuver can be performed. When a unifying theory is first proposed, one can only guess that it will eventually allow us to perform the Friedman maneuver. Circumstances may turn out otherwise—the unifying theory may not receive the independent confirmation that was hoped for. The difference between the two maneuvers is thus captured to some extent by the difference between conceptual invention and empirical discovery: one invents a theory that is tailor-made for the Bayesian maneuver, but one discovers by research that a theory is suitable for the Friedman maneuver.

There is a third scenario belonging to the same category of theoretical strategies that is based neither on a conceptual invention nor on an empirical discovery, but on a conceptual discovery. Suppose that theories T_1 and T_2 have both been around for a long time, and that they are believed to be logically independent. Suppose also that, as a result of their empirical track record, T_1 has been assigned a greater probability than T_2. Now suppose that it's discovered that T_1 entails T_2. This conceptual discovery shows that our probability function has been incoherent all along, though we didn't know it. Now that we do know it, of course, we have to alter our probabilities in such a way that the incoherence is eliminated. The general situation is this: we discover a new logical truth about our hypotheses that reveals an unsuspected incoherence, whereupon we must change our opinion about the plausibility of at least some of these hypotheses. I've given the name of *amplification* to this sequence of events (Kukla 1990a). Amplification is a third type of theoretical maneuver relating to evolving probability. It's like the Friedman maneuver in requiring a discovery rather than an invention. It's like the Bayesian maneuver in not requiring any new empirical research.

A paradigmatic example of amplification is to be found in the famous debate between Chomsky and Putnam over the innateness of universal grammar. The basic Chomskian argument for innateness is well known. Chomsky (1980) notes that linguistic data so drastically underdetermine the grammar that it's impossible for language learners to hit upon the correct grammar ab ovo by any rational procedure. His conclusion is that the language learner's search through the solution space of possible grammars must be innately constrained to a finite subset of possible grammars. This is the *innateness hypothesis*. Chomsky also notes that the innateness hypothesis also provides an explanation for the existence of linguistic universals (grammatical features common to all human languages). Another way to explain linguistic universals is to hypothesize that all human languages are descendants of one and the same ancestral language. This is the *common origin* hypothesis. Chomsky expresses the view that the common origin hypothesis is extremely implausible (235). Putnam (1980), however, has a clever argument to the effect that the common origin hypothesis is *entailed* by the innateness hypothesis (conjoined with other features of Chomsky's views):

> Suppose that language-using human beings evolved *independently* in two or more places. Then, if Chomsky were *right*, there should be two or more *types* of human beings descended from the two or more original populations, and normal children of each type should fail to learn the languages spoken by the other types. Since we do not observe this . . . we have to conclude (if the [innateness hypothesis] is true) that language-using is an evolutionary "leap" that occurred only *once*. But in that case, it is overwhelmingly likely that all human . . . languages are descended from a single original language. (246)

Thus, the innateness hypothesis, together with Chomsky's hypothesis that it's impossible to learn a language ab ovo and the observation that every child can learn any language, *entails* the common origin hypothesis. But then it's incoherent to suppose that the former is plausible while the latter is implausible. Putnam's argument forces Chomsky to make a theoretical adjustment: he must grant either that the common origin hypothesis is "overwhelmingly likely" or that some feature of his own view is overwhelmingly unlikely.

The discovery of a hitherto unsuspected relation of entailment between two theories that were thought to be probabilistically independent *always* effects an amplification, even if the probability of the antecedent theory wasn't incoherently greater than the probability of the consequent theory. For if the theories were thought to be independent, it will have been true that $p(T_1 \& T_2) = p(T_1)p(T_2)$. After the discovery that T_1 entails T_2, however, we have $p(T_1 \& T_2) = p(T_1)$. At the very least, the discovery of an unsuspected entailment between two previously unconnected theories makes their conjunction more probable.

All the examples of amplification that we've seen so far have involved the entailment of one theory by another. But it's clear that there are indefinitely many logical properties and relations that can play the same role. For instance, the discovery that a theory is internally inconsistent counts as a type of amplification, for it's a logical discovery that reveals a preexisting incoherence. Another type of amplification occurs when two extant theories are found to be inconsistent with each other. If the theories were previously both thought to be plausible, that is, to have probabilities greater than 0.5, then the discovery that they're mutually inconsistent once again reveals that our probability function was incoherent. The result, once again, is that we're forced to change our mind: either one theory or the other (or both) must no longer be regarded as plausible. There are analogous operations in the Bayesian and Friedman maneuvers. For example, a theoretician may set out to discredit a "received" theory T_1 by constructing a highly implausible T_2 that is entailed by T_1 (equivalently, by constructing a highly plausible *not*-T_2 that is inconsistent with T_1). If this enterprise succeeds, the received theory must diminish in plausibility. I would consider the whole operation to be an instance of the Bayesian maneuver. Similarly, the following procedure can be viewed as a variant on the Friedman maneuver: theory T_1, which is incompatible with T_2, receives continued confirmation, as a result of which $p(T_1)$ continues to rise. At the point where $p(T_1) + p(T_2)$ reaches unity, however, our probability function threatens to become incoherent. If T_1 receives any more confirmation beyond this point, we must begin to decrease the probability of T_2. This is, of course, a common event in the history of science. It occurs whenever two incompatible hypotheses are subjected to empirical testing and one of them wins. The location of both this commonplace scientific practice and theoretical unification à la Friedman in the same place in our conceptual framework reveals their considerable kinship and, I think, goes a long way toward reducing any sense of strangeness that might attach to Friedman's proposal.

4.5 Laudan on Conceptual Problems

The several theoretical strategies I've discussed have implications for Laudan's (1977) analysis of the varieties of *conceptual problems* in science.[2] Laudan distinguishes these from *empirical problems*. The latter are first-order questions about the entities in some domain, while the former are higher order questions about the conceptual systems devised to solve the first-order questions. According to Laudan, a piece of scientific work constitutes a scientific advance if it results in an increase in the difference between the number of empirical problems solved and the number of conceptual problems generated by these solutions. The class of conceptual problems is subdivided into *internal* and *external* conceptual problems. Internal problems include internal inconsistency, conceptual ambiguity, and circularity. External conceptual problems arise as a result of a theory's "cognitive relations" to other scientific theories, to methodological principles, or indeed to any aspect of a prevalent worldview. Laudan distinguishes the following subvarieties of external conceptual problems: (1) inconsistency (a theory T_1 entails the negation of another theory T_2), (2) implausibility (T_1 entails that T_2 is unlikely), (3) compatibility (T_1 entails nothing about T_2 when we have reason to expect at least a reinforcing connection), and (4) reinforcement (T_1 renders T_2 merely more likely when we have reason to expect full entailment). According to Laudan, "any relation [between two theories] short of full entailment could be regarded as posing a conceptual problem for the theories exhibiting it," although they pose very different degrees of cognitive threat (54). The degrees of cognitive threat are represented, in descending order, by (1)–(4).

There are obvious correspondences between this notion of conceptual problems and the class of theoretical strategies I have described in section 4.4. Consider an amplification in which two highly plausible theories T_1 and T_2 are shown to be mutually inconsistent. In Laudan's scheme, we would say that such a maneuver creates a conceptual problem for the theories involved. More generally, the resolution of a Laudanian conceptual problem is a piece of theoretical work of the type that I've delineated. Laudan's account of this type of episode strikes me as eccentric, however. The way he pictures it, the contribution to science comes only when we resolve the incoherence by adjusting our opinions of the theories' merits. The original *discovery* that necessitated this reappraisal is not itself a contribution to the progress of science. Indeed, on Laudan's account (progress equals empirical problems solved minus conceptual problems created), we would have to consider the logical discovery to be *retrogressive*. But then why would anyone ever undertake to demonstrate an inconsistency between two theories? Even if the inconsistency is eventually resolved, we would merely have subtracted one unit of progress and added it back again. Since the creation and solution of conceptual problems leaves us where we started, it's difficult to understand why anyone would *look* for relationships like inconsistency. Laudan's scheme fails to recognize that the *creation* of conceptual problems may also be a scientific advance.

I think that Laudan could accommodate the foregoing criticism with very minor changes in his account of theoretical work. A more serious problem is the fact that his discussion of the cognitive relations that produce cognitive problems fails to take account of the all-important factor of plausibility. Consider his assertion that the relation of full entailment between two theories is never problematic. If we believe both T_1 and T_2 to be true, then it is indeed the case that the discovery that T_1 entails T_2 poses no cognitive threat. But suppose we are adherents to T_1 who regard T_2 as a discredited rival. In that case, the discovery that T_1 entails T_2 would create a severe conceptual problem for T_1. Similarly, Laudan's ranking of theoretical relations in terms of their degree of cognitive threat makes sense only under the assumption that we're partisans of both theories in the relationship. In other circumstances, the discovery that T_1 is incompatible with T_2 might not be a problem at all — as when T_1 is a theory that we like and T_2 is its discredited rival. It isn't possible to identify conceptual problems solely on the basis of the cognitive relationships that Laudan discusses. At the very least, some notion of plausibility or acceptability has to be brought into the analysis.

The phenomena of evolving probability point to a still more serious deficiency in Laudan's scheme. Consider the following scenario: T_1 and T_2 are well-established theories that are thought to be independent, but it's shown that T_1 entails T_2. By Laudan's formula for scientific progress (empirical problems solved minus conceptual problems generated), nothing has happened here at all. No new empirical problems are going to be solved as a result of this logical discovery, nor are any conceptual problems either created or resolved. But this discovery is clearly to be counted as a scientific advance. At the very least, it makes the *conjunction* of T_1 and T_2 more plausible than it was before (and thus it qualifies as an amplification). In this case, the scientific advance produced by a conceptual analysis of theories results in an increase in conceptual goodness rather than a decrease in conceptual badness. Laudan's analysis makes no provisions for such conceptual improvements. His account of theoretical work in science commits him to the view that the only legitimate kinds of theoretical work are (1) the construction of theories that solve new empirical problems, and (2) the repair of conceptual deficiencies. But amplification and the Bayesian maneuver fit into neither category. Rather, they are theoretical strategies for accomplishing the end that experimentalists pursue when they submit a theory to empirical test — namely, altering the epistemic standing of a hypothesis. As in the case of empirical testing, the outcome may be either an elevation or a decrease in that standing. Evolving probability is the extension of theory testing to the conceptual arena.

Realism and
Underdetermination I

*Does Every Theory Have Empirically
Equivalent Rivals?*

The main argument for antirealism is undoubtedly the argument from the underdetermination of theory by all possible data. Here is one way to represent it: (1) all theories have indefinitely many empirically equivalent rivals; (2) empirically equivalent hypotheses are equally believable; (3) therefore, belief in any theory must be arbitrary and unfounded.

It's obviously unprofitable to question the deductive validity of the argument from underdetermination. This leaves the realist with two broad lines of defense: to claim that one of the argument's premises is incoherent, or that one of them is false. Among recent incoherence claims have been attacks on the distinction between empirical and theoretical consequences (e.g., Churchland 1985) and on the distinction between believing a theory to be true and the supposedly weaker epistemic attitudes recommended by antirealists (e.g., Horwich 1991). These important issues are dealt with in later chapters. In chapters 5–7, it is a presupposition of the discussion that the premises of the argument are coherent. Let's call the first premise of the argument EE (for "empirical equivalence") and the conclusion UD (for "underdetermination"). Chapter 5 is devoted to an exploration of the status of EE. The second premise—the thesis that EE entails UD—is taken up in chapters 6 and 7.

After some clarification of the argument in section 5.1, I consider three current lines of attack on EE. The first, and most audacious, is the claim that EE is demonstrably false—that it can be shown that any two hypotheses are empirically discriminable in some possible state of science. This argument is dealt with—and repudiated—in section 5.2.

Assuming that the disproof of EE fails, realists may fall back on the weaker claim that the truth of EE has not been established. Note that to establish EE, it isn't sufficient to show that some theories—or even all extant theories—have empirically equivalent rivals. Realists need not be committed to the view that any of our *current* theories ought to be believed. To be a (minimal epistemic) realist, one needs only to maintain that there are *possible* states of science wherein it's rational to hold theoretical beliefs. For the underdetermination argument to undermine this realist claim, it needs to be established that there are empirically equivalent rivals to *any* theory, including the theories of the future whose content we are presently unable to specify. Clearly, the only way to establish this proposition is to provide a universal algorithm for constructing rivals from any given theory.

Such algorithms have been offered by antirealists. Van Fraassen's favorite is this one: given any theory T, construct the rival T' which asserts that the empirical consequences of T are true, but that T itself is false (see e.g., van Fraassen, 1983). T and T' are empirically equivalent by definition. Moreover, T and T' can't both be true, since they contradict each other. This construction is offered by van Fraassen as a deductive proof of EE. However, several writers have rejected this proof on the following grounds: even though T' is both empirically equivalent to T and incompatible with T, it suffers from the critical shortcoming of failing to be a scientifically acceptable rival hypothesis; hence, it doesn't count toward the thesis that there are empirically equivalent rivals to every theory T. The same claim has been made regarding other proposed algorithms for producing empirically equivalent rivals. Some general features of this argumentative strategy are discussed in section 5.3. In sections 5.4–5.7, I investigate several criteria for what a genuine rival must be like. My conclusion is that these putative requirements are epistemically uncompelling and that they fail to invalidate one or another of the available algorithms for producing empirical equivalents to T.

The third line of attack allows that there may be algorithms for producing empirically equivalent formulations, and that these formulations may in fact be genuine rivals. However, it is claimed, the competition must inevitably be lost by the algorithmic constructions. This type of argument evidently calls on us to reject the second premise of the underdetermination argument, which says that EE entails UD. The status of this premise isn't at issue in this chapter. Indeed, unless otherwise indicated, I will assume for the sake of the argument that the second premise is true. However, in section 5.8, I argue that even if the second premise is accepted, the claim of epistemic inferiority for the algorithm-produced rivals of T has no discernable support. The overall conclusion, in section 5.9, is that there are powerful reasons for accepting EE as true. The fate of the argument from underdetermination thus hangs on the status of the claim that EE entails UD.

I noted earlier that van Fraassen has arguments designed to bolster the case for EE. In addition to the algorithmic construction T', he also devotes a section of *The Scientific Image* to developing the thesis that Newtonian mechanics has indefinitely many empirically equivalent rivals that recognizably have the form of traditional theories. Moreover, I show in chapter 7 that van Fraassen has also been the foremost defender of the second premise of the argument for UD—the claim that empirically equivalent theories are epistemically equivalent. It's natural to suppose that these philosophical labors have been undertaken in the service of the underdetermination argument. Yet it's curiously difficult to locate the exact place in van Fraassen's writings where this argument is presented in fully general form. All the pieces of a defense of the argument are there to be found. But there's no place where they're all pulled together. Worrall, writing in 1984, says that the argument from underdetermination is "mentioned" in *The Scientific Image*, but that it isn't systematically developed. Since then, van Fraassen has provided us with additional pieces of an argument, but I think that Worrall's judgment on this matter still stands.

5.1 The Varieties of Arguments from Underdetermination

Let's suppose that we have an algorithm, or a number of algorithms, for producing acceptable rivals to any theory T. It doesn't yet follow that there are, as EE stipulates, *indefinitely many* rivals to T. What is the effect on the argument from underdetermination of weakening premise EE to the assertion that there exist *some* empirically equivalent rivals to any T? Hoefer and Rosenberg (1994) claim that this would already be enough to warrant the rejection of realism. They repudiate Quine's (1975) suggestion that in such a case "we may simply rest with both systems and discourse freely in both" (328), on the grounds that such discourse would violate the law of the excluded middle. And so it would, if we merely asserted both hypotheses. Moreover, I agree with Hoefer and Rosenberg that "Quine is [also] wrong if he means to suggest that one explores and assesses alternative hypotheses by oscillating in belief between them" (606). But there are more charitable — and less problematic — ways of interpreting Quine's recommendation. Let T1 and T2 be empirically equivalent and mutually incompatible rivals, both of which are maximally supported by the data. Then we are free to assert that T1 or any of its consequences is true *on the condition that T2 is false*, and vice versa. This is a full and unproblematic expression of the state of our knowledge in this situation. Our ability to "discourse freely about both" comes from the freedom of saying everything that either T1 or T2 says, as long as it is always understood that the negation of the rival is a presupposition of the discussion. The question is: is this realism?

Well, suppose that T1 posits the existence of unobservable A-particles, that T2 posits unobservable B-particles, and that A-particles and B-particles have incompatible properties, in the sense that the existence of either type entails the nonexistence of the other type. Granting that the second premise of the underdetermination argument is true (I won't continue to mention this proviso), it follows that we will never be able to establish whether A-particles or B-particles exist. But it remains possible for us to learn that *AB*-particles exist, where an AB-particle is defined as an entity that is either an A-particle or a B-particle. And isn't it the case that AB-particles are theoretical entities? If they are, we may be able to obtain sufficient warrant for believing in theoretical entities even if every theory has empirically equivalent rivals. In fact, we can even establish what some of their properties are: if A-particles have property P and B-particles have property Q, then we may hope to learn that there exist unobservable entities that have the property P-or-Q. Doesn't this satisfy the requirements for minimal epistemic realism?

There are, I know, some philosophers who are unmoved by artificial constructions of this type. These philosophers would be tempted to say that conceptual monstrosities like AB-particles don't qualify as bona fide theoretical entities. But even if this is so, it still has to be admitted that believers in T1–or-T2 are justified in believing *that theoretical entities exist*. AB-particles may not be bona fide theoretical entities, but A- and B-particles are; and if T1–or-T2 is true, then either A-particles exist or B-particles exist. In either case, it follows that there are theoretical entities. Is this realism? Hoefer and Rosenberg evidently think not. Leplin has recently expressed the same view:

The situation is rather like a criminal investigation, in which the evidence shows that either the butler or the maid is guilty but does not indicate which. Then no one can be convicted. If empirical equivalence is rampant, it would seem that we cannot be realists because we would never know what to be realists *about*. (1997, 154)

But is it obviously the case that we can't be realists if we don't know what to be realists about? Our discussion reveals an ambiguity in the concept of scientific realism. To be a realist is to think that there are circumstances in which it's rational to believe in theoretical entities. But to "believe in theoretical entities" can be interpreted as (1) believing that theoretical entities exist, or (2) there being theoretical entities X such that one believes Xs exist. Which doctrine is the real realism? There's no point haggling over labels. Let's call the first doctrine *abstract realism* and the second *concrete realism*. Now suppose that it's established that for any T, there are finitely many empirically equivalent rivals to T that posit different theoretical entities. If one accepts entities like AB-particles as bona fide theoretical entities, then one's realism isn't the least bit affected by this discovery. If, on the other hand, one refuses to take AB-particles seriously, then the same discovery forces us to abandon concrete realism, but it leaves abstract realism untouched. There's no need to settle the status of AB-particles here. Whatever we say about them, it's clear that the existence of finitely many rivals that posit theoretical entities fails to rule out states of opinion that are stronger than any that have ever been called antirealist.

What if we have an algorithm for constructing rivals to any T that do *not* postulate theoretical entities? T', as defined above, is such a construction. In this case, the existence of a single rival is already enough to refute abstract realism as well as concrete realism. But there remains a state of belief that concedes more to realism than any extant antirealism, and that is not ruled out by the existence of a single rival, or any finite number of rivals, that do not postulate theoretical entities. Suppose that $T_1, T_2, \ldots,$ T_n are such rivals for any T. Then there's nothing in the underdetermination argument itself that would disallow us from splitting the epistemic difference among these hypotheses and saying that T and the T_i are equiprobable. But to say that T and the T_i are equiprobable is to admit that there's a nonzero probability that T is true, which in turn entails that there's a nonzero probability that the theoretical entities posited by T exist. This is a greater concession to realism than van Fraassen or any other antirealist has been willing to make. But it's also weaker than concrete realism—for the admission that T and the T_i are empirically equivalent (and that EE entails UD) entails that even if T's empirical consequences are maximally confirmed, $p(T)$ can never rise above any of the $p(T_i)$—that is, that $p(T)$ can never rise above the value $1/(n + 1)$. Let's call the doctrine that there are circumstances wherein it's rational to ascribe a nonzero probability to the existence of theoretical entities, together with the admission that these nonzero probabilities can never be elevated beyond a ceiling value that is too low to warrant full belief, by the name of *feeble* realism. There is a feeble abstract realism as well as a feeble concrete realism.

Contra Hoefer and Rosenberg, the realist's view of the world is more severely challenged by the construction of infinitely many than by finitely many rivals, whether or not these rivals posit theoretical entities. An algorithm that provides infinitely many rivals that posit theoretical entities refutes feeble as well as full-fledged concrete realism (though it remains compatible with full-fledged abstract realism), and an algorithm that provides infinitely many candidates that do *not* posit theoretical entities refutes even feeble abstract realism.

The upshot is that there are several different EEs and correspondingly several different arguments from underdetermination, the soundness of which would refute several different grades of scientific realism. After all the arguments have been presented, we have to ascertain just which forms of EE have been sustained and which varieties of realism are refuted by these EEs. In the intervening sections, I usually allow the context to determine which of the several EEs is at issue.

5.2 The Arguments from the Variability of the Range of the Observable and the Instability of Auxiliary Assumptions

Laudan and Leplin (1991) give two arguments that purport to show that there can be no guarantee that any theory has empirically equivalent rivals. According to the argument from "the variability of the range of the observable", what counts as an observable phenomenon undergoes historical changes with the development of new observational technologies. Prior to the invention of the microscope, microorganisms were not observable; now they are. Thus, even if two theories are currently empirically equivalent, there can be no guarantee that they'll continue to be so in the future.

One problem with this argument is that it employs a notion of observability that is explicitly repudiated by most contemporary antirealists. For Laudan and Leplin, observability is roughly the same thing as detectability by any means available. There's no doubt that we're now able to detect phenomena that were previously undetectable, and that currently undetectable phenomena may become detectable in the future. But for this very reason, van Fraassen (1980) rejects detectability *tout court* as a criterion for observability, and adopts instead the criterion of "detectability by the unaided senses". I examine this van Fraassian idea in detail in chapter 10. At the present juncture, it suffices to note that Laudan and Leplin's first argument knocks down a straw person. Changes in observational technology may quite reasonably be said to alter what we deem to be observable—but not in van Fraassen's sense of the term. I offer another criticism of this argument later in this section.

The argument from "the instability of auxiliary assumptions" was anticipated by Ellis (1985). In this critique, I follow Laudan and Leplin's more extensive presentation. The argument begins by noting that the empirical consequences of a theory depend on which auxiliary theories are permitted in making derivations. But like the "range of the observable", the permitted auxiliaries are liable to historical change. Therefore, given any two theories that are logically distinct and presently thought to be empirically equivalent, there is a possible state of *future* science in which they become empirically distinct. And therefore theories are not guaranteed to have empirical equivalents.

In an earlier essay (Kukla 1993), I criticized Laudan and Leplin's argument as follows. Suppose that I believe that for every theory T1, there is an empirically equivalent theory T2, and that I'm confronted with their argument. The fact that the auxiliaries can change might impel me to reconstrue the notion of empirical equivalence as a relation between *indexed* theories — that is, couples consisting of a theory and a specification of the permissible auxiliaries. But there's nothing in the argument that would force me to give up the view that every *indexed* theory has empirically equivalent rivals with the same index.

Leplin and Laudan have responded to this counterargument as follows:

> EE [the thesis of empirical equivalence] is clearly intended, by proponents and detractors alike, as an atemporal thesis. . . . It denies for any theory the possibility of (ever) observationally discriminating it from some rival theory. . . . This suggestion [that every indexed theory has an empirically equivalent theory with the same index] . . . does nothing to undermine our argument against EE. . . . Empirical equivalence is best not thought of as a condition that can be maintained temporarily; that is the point of using "equivalence", a formal term. So let us put the point in terms of observational indiscriminability. EE stands refuted if, to retain observational indiscriminability, we are forced to have recourse to changing rivals. For EE does not (merely) assert that at all times there is some theory from which a given one is indiscriminable. Rather, it asserts that for any theory there is some rival from which it is unqualifiedly (in principle) indiscriminable (i.e., to which it is empirically equivalent). (1993, 8–9)

This seems to me to be an inadequate rebuttal. Let A*t* be the permitted auxiliaries at time *t*. It's true, as Leplin and Laudan note, that the fact that the indexed theory (T1, A*t*) has the same observational consequences as (T2, A*t*) doesn't mean that T1 is atemporally empirically equivalent to T2. Empirical equivalence at *t* is not the same thing as empirical equivalence *tout court*. But the fact that (T1, A*t*) is empirically equivalent to (T2, A*t*) means that *we believe at time t* that T1 and T2 are atemporally empirically equivalent (since we believe at *t* that the auxiliaries A*t* are correct). The claim is not merely "that at all times there is some theory from which a given one is indiscriminable". It's just what Leplin and Laudan require–that at any time *t*, we have reason to believe that "for any theory there is some rival from which it is unqualifiedly (in principle) indiscriminable." To be sure, our opinion as to what these timeless rivals are will change with the auxiliaries. But the point is that we know that, whatever our future opinion about the auxiliaries will be, there will be timeless rivals to any theory under those auxiliaries.

Here's an alternative presentation of the same point that doesn't depend on an intricate ballet of quantifiers. The inspiration comes from a refutation by Boyd of an ancestor of Laudan and Leplin's argument:

> It is universally acknowledged that in theory testing we are permitted to use various well-confirmed theories as "auxiliary hypotheses" in the derivation of testable predictions. Thus, two different theories might be empirically equivalent . . . but it might be easy to design a crucial experiment for deciding between the theories if one could find a set of suitable auxiliary hypotheses such that when they were brought into play as additional premises, the theories (so expanded) were no longer empirically equivalent. (Boyd 1984, 50)

Boyd's reply is that the thesis of empirical equivalence can simply be reformulated so that it applies not to individual theories, but to "total sciences", that is, the conjunction of all our acceptable scientific theories:

The thesis, so understood, then asserts that empirically equivalent total sciences are evidentially indistinguishable. Since total sciences are self-contained with respect to auxiliary hypotheses, the rebuttal we have been considering does not apply. (50)

Laudan and Leplin's argument differs from its ancestor by introducing a temporal dimension: even if T_1 and T_2 have the same empirical consequences under the *current* auxiliaries, this fact would not establish their timeless empirical equivalence, since there are possible states of future science in which the new auxiliaries permit an empirical discrimination between them to be made. But the new argument is just as susceptible as the original to Boyd's countermove. Let Tt be the total science that we subscribe to at time t. Then Laudan and Leplin's point that the permissible auxiliaries may change amounts to the claim that our total science may change—that for any other time t', total science Tt' may not be identical to total science Tt. Now, the argument from the instability of auxiliaries doesn't refute the claim that for every indexed theory (i.e., theory *cum* auxiliaries), there is an empirically equivalent theory with the same index—nor do Leplin and Laudan claim that it does. Their point is that this assumption about indexed theories doesn't establish EE. But a total science is nothing more or less than the conjunction of any "partial" theory and all the auxiliary theories that we deem to be permissible. It doesn't matter which partial theory we begin with—the end result will be the same. Thus, the assumption that every indexed theory has an empirical equivalent with the same index is logically equivalent to the claim that for every total science, there is another total science that is *timelessly* empirically equivalent to it, just as Leplin and Laudan require. Evidently, their argument fails to show that "we are forced to have recourse to changing rivals" (1993, 9) to provide empirical equivalents to total sciences: the empirically equivalent rival to Tt remains so for all time. It's true that our current total science is bound to give way to a different one, and thus, it's also true that the existence of empirical equivalents to Tt will no longer serve to establish that the future total science Tt' has empirical equivalents. But, of course, proponents of EE never believed otherwise. It was always understood that new theories would arise that would require us to construct new empirical equivalents. Belief in EE is the belief that, whatever the new theories may be, it will always be possible to find (eternal) empirical equivalents to them. The most that Leplin and Laudan's argument about auxiliaries can possibly establish is that *partial* theories need not have empirical equivalents. Whether we count this as a refutation of EE is an unimportant terminological issue. What matters is that the existence of timeless empirical equivalents to total sciences brings in its train all the epistemological problems that were ever ascribed to the doctrine of EE.

The move to total sciences also provides another reason to reject the argument from the variability of the range of the observable. The first problem with that argument was that it employs a notion of observability that is explicitly repudiated by contemporary antirealists like van Fraassen. The second problem is that even regarding its own notion of observability, the argument doesn't apply to total sciences. For let P be a phenomenon that is referred to by Tt, our total science at time t, and suppose that P is deemed to be unobservable at time t. Since Tt is our *total* theory about the world, the claim that P is unobservable should be a *part* of Tt. When, at a later time t', P comes to be counted among the observables, this change will require us to alter our total science — for T*t'*, our total science at *t'*, must contain the claim that P *is* observable. Thus, changes in observables, like changes of auxiliaries, can be assimilated to changes in total sciences.

Leplin (1997) has recently presented an alternative and more elegant formulation of the argument from the instability of auxiliaries. This argument purports to show (roughly) that UD entails the negation of EE. If this is true, then one premise or the other of the underdetermination argument must be false. For suppose that both premises are true — that EE is true and that EE entails UD. Then the conclusion UD follows. But if Leplin is right in his claim that UD entails not-EE, it further follows from UD that EE is false, which contradicts an assumption that we began with. Therefore, both premises can't be true. This argument doesn't tell us which of the premises is at fault — it merely tells us that at least one of them is false. Thus, it belongs neither in chapter 5, which deals with refutations of the first premise, nor in chapter 6, which deals with refutations of the second premise. I place this discussion here because it's too short to warrant a chapter of its own.

Leplin's argument runs as follows. Let T be an arbitrary theory, and suppose that UD is true — that is, that for every theory, there are incompatible rival theories that have just as much warrant to be believed. Now, the empirical content of T is identical to the empirical consequences of the conjunction of T and the currently acceptable auxiliaries, A. But by UD, there's a rival set of auxiliaries A' that have just as much warrant as A. Thus, the empirical content of T is indeterminate between the empirical consequences of T & A and those of T & A'. But if the empirical content of individual theories is indeterminate, then so is the truth of EE, which claims that for every theory, there are others with the same empirical content. The conclusion is that the argument from EE to UD cannot be sound: either EE is rationally unbelievable, or it doesn't lead (in conjunction with other believable premises) to UD.

If Leplin's argument itself is sound, it accomplishes everything that Laudan and Leplin had hoped to accomplish with their attempted refutation of EE, and it does so with a greater economy of means. It isn't necessary to show that EE is false in order to defuse the argument from underdetermination. It suffices to show that one can't accept both EE and UD. If we don't accept EE, then one of the crucial premises of the underdetermination argument is missing, and if we *do* accept EE, then, according to Leplin's argument, we are required to repudiate UD. Either way, the argument from underdetermination fails.

But this new and improved argument has the same loophole as the original: it doesn't apply to total sciences. If T is a total science, then its empirical content doesn't depend on any auxiliary hypotheses. Therefore, the empirical content of T can be determinate even if UD is true. Antirealists have the option of endorsing a version of EE which asserts only that for every total science, there are empirically equivalent rival total sciences. The argument from this version of EE to the conclusion UD circumvents the difficulty that Leplin brings up. To be sure, the conclusion won't apply to "partial" theories. Leplin's argument shows that antirealists who rely on the underdetermination argument should indeed regard the empirical content of partial theories as indeterminate. But the underdetermination of total sciences hasn't been refuted, and this principle, conjoined with the admission that partial theories have no determinate empirical content, constitutes an antirealist stance that is untroubled by Leplin's point.

Realists will undoubtedly wish to attack the notion of a total science. Admittedly, this notion has been severely underanalyzed by both friends and foes of UD. But it remains to be seen whether its obscurities affect the role it plays in the argument for UD. The prima facie case has been stated. The burden of proof is on realists to show why the total sciences version of the underdetermination argument fails.

5.3 The Concept of a "Genuine Rival"

Of course, the refutation of the argument from the instability of auxiliaries doesn't establish that EE is true. Laudan and Leplin claim that it's "widely supposed that a perfectly general proof is available for the thesis that there are always empirically equivalent rivals to any successful theory" (1991, 449), but that this received view is without foundation. Similarly, Hoefer and Rosenberg write that "no general arguments show that two empirically equivalent total theories are possible or impossible (or that they are inevitable)" (1994, 601). I think that both pairs of authors, HR and LL for short, are wrong on this point. The curious thing about their claim is that both HR and LL mention algorithms that seem to produce empirically equivalent rivals to any theory. We've seen one of them already: for any theory T, construct the rival T' which asserts that the empirical consequences of T are true but that T itself is false. Both HR and LL dismiss this construction (and others) as trivial and as failing provide a genuine rival to T. For LL, T' is an species of "logico-semantic trickery" (1991, 463); for HR, it's a "cheap trick" (1994, 603). Both seem to concede that T and T' are empirically equivalent. At least regarding to total sciences, it's difficult to see how one can avoid making this concession. But apparently, both wish to make the claim that some empirical equivalents to T are not to be counted among the rivals for the purpose of assessing the truth of EE, the hypothesis that every theory has empirically equivalent rivals.

In sections 5.4–5.7, I evaluate specific proposals for why T' and kindred constructions should be discounted in this way. But first, I wish to discuss some general features of the philosophical strategy of discounting trick hypotheses. The account that LL and HR have in mind seems to go roughly as follows. Even though a hypothesis may possess the traditional scientific virtues of having a truth-value, being confirmable and disconfirmable, and generating indefinitely many testable predictions, it might nevertheless be excluded from serious scientific discourse for failing to satisfy an a priori constraint on the proper form for a scientific hypothesis. Foregoing any originality, let's call this missing property by the name of *charm*. A hypothesis like T' that has many of the traditional virtues of scientific hypotheses but that (purportedly) fails to be charming can be called a *quasi-hypothesis*. Since T' is merely a quasi-hypothesis, it's irrelevant to the question whether there are any *genuine scientific hypotheses* that are empirically equivalent to T.

My first reaction to this argument is: hypothesis, schmypothesis. Since T' is truth-valuable, it can't be denied that T' is an appropriate candidate for belief or disbelief. Thus, it's at least conceptually possible that we might have better grounds for believing an algorithmically produced quasi-hypothesis than for believing the corresponding hypothesis. Indeed, Hoefer and Rosenberg (1994) tacitly admit that quasi-hypotheses may actually be the only available way to describe the truth. On the one hand, they cite "almanac-style lists of observation conditionals" as prime candidates for quasi-hypothesishood (603). On the other hand, they make the point that the universe may not be describable in any way other than an almanac-style list:

> There is . . . no guarantee that genuine empirical equivalence of adequate total theories will arise. A further factual precondition must be met: The world must admit of at least one empirically adequate total theory. That this is the case is usu-ally taken for granted; but in the absence of a compelling metaphysical argument that such a total theory must be available, it should not be assumed (and empiri-cists, of course, cannot allow such metaphysical arguments). Of course, an alma-nac-style listing of observation conditionals is always in principle possible, but . . . such a bizarre construction need not be counted as a genuine theory. (604)

In light of the fact that a "bizarre construction" may express the truth—in fact, that it may express as much of the truth as we are able to express—what, exactly, is the import of not counting it as a genuine hypothesis? Withholding the honorific title from almanac-style listings or from T' merely forces proponents of the underdetermination argument to recast the argument in terms of *schmypotheses*, a schmypothesis being anything that is either a genuine hypothesis or a quasi-hypothesis. The new argument would look just like the original, except that "schmypothesis" would appear wherever "hypothesis" occurred in the original. Even LL and HR would have to accede to the first premise of the revised argument—namely, that every schmypothesis has empirically equivalent rival schmypotheses. The verdict on the argument from underdetermination would depend entirely on the status of the *second* premise—that empirically equivalent schmypotheses are equally believable. This does not, of course, mean that the dismissal of T' is a mistake. But it does suggest that the dialectical situation produced by the tacit appeal to the charm of hypotheses is a fatiguingly familiar one. The new claim seems to be of a piece with traditional assaults on the *second* premise of the argument for UD with claims that explanatoriness, or simplicity, or some other nonempirical property of theories has epistemic import. I examine these arguments in chapter 6, and I come to the conclusion that all arguments of this form engage in question-begging.

Be that as it may, the essays by HR and LL are intended to be discussions of the *first* premise of the underdetermination argument, EE. The fact that both pairs of authors regard the repudiation of T' as relevant to the first premise indicates that they view the deficiencies of T' in a different light from those of theories that offer bad explanations, or are unnecessarily complex. For these authors, T' doesn't merely lose to T in an evaluation of their relative epistemic merits—T' *isn't even in the running*. It is not a candidate for epistemic evaluation. This distinction between (1) a proposi-tion being evaluated and found to be seriously deficient in epistemic virtue and (2) a proposition being unworthy of epistemic evaluation strikes me as obscure, perhaps incoherent. I can understand the distinction when it's applied to propositions ver-sus, say, apples: an apple doesn't score low marks on an epistemic evaluation—it's not even a candidate. But when the putative candidate is a truth-valuable claim about the world, in what sense is it not amenable to epistemic evaluation? T' is at least a

logically possible state of affairs. In fact, we've seen that it's logically possible that T'
is the *whole* truth about the universe. I don't wish to deny that there are excellent
reasons for doubting that any hypothesis having the form of T' is the whole truth about
the universe. What I am denying is that there's a coherent sense in which T' is a
noncandidate for epistemic evaluation, like an apple, because of some feature of its
form or its relation to T.

5.4 The Parasitism Criterion

But let us, for the sake of the argument, grant that such a priori dismissals are permissible. The question remains: why make T' the victim? What is it about T' and
other algorithmically constructed rivals that disqualifies them for any career, whether
illustrious or ignoble, as a bona fide scientific hypothesis? Neither LL in 1991 nor
HR in 1994 discuss specific criteria for charming hypotheses. Indeed, LL's rhetoric
suggests that this is not a serious subject for philosophical discussion. I think it is,
quite aside from its bearing on the underdetermination issue. Let's agree that the
algorithmic rivals are too absurd for anyone to take seriously as total theories about
the world. It's a fallacy to infer from this that there can be no value in elucidating the
means whereby such judgments are made—the same fallacy as supposing that the
psychology of silliness must be a silly subject. For one thing, the artificial intelligence
project of constructing an artificial scientist could not succeed unless our program
provided resources for *identifying* quasi-hypotheses as such. HR, for their part, concede that criteria for charming hypotheses need to be formulated. They even admit
that the selection of such criteria is a matter of some delicacy:

> While we do not want to admit such formulations [e.g., T'] as epistemologically
> significant rivals, at the same time we do not want to adopt criteria for their exclu-
> sion that go too far and have the potential of discarding a rival that is both truly
> distinct, and possibly true. (604)

Nevertheless, like LL in their 1991 essay, HR give us no clue as to what these criteria
may be.

I complained about this dereliction in my 1993 critique of LL's essay (HR's equally
derelict essay had not yet been published). In their response, Leplin and Laudan
undertook to provide "the rudiments of an analysis" of the concept of charm (1993,
13), which I was then calling "theoreticity". The theme of this and the next section is
that these rudiments need more work. The central element of their analysis is the
claim that T' is quasi hypothetical on the basis of what may be called the *parasitism*
criterion:

> T' is totally parasitic on the explanatory and predictive mechanisms of T. . . . a [real]
> theory posits a physical structure in terms of which an independently circumscribed
> range of phenomena is explainable and predictable. (13)

The emphasis here, I think, must be on the independence of the formulation rather
than on the explanatoriness, for surely Leplin and Laudan do not want to rid science
of the nonexplanatory phenomenological laws of physics. I interpret their claim as

follows: a putative hypothesis is merely a quasi-hypothesis if its formulation necessarily involves a reference to another hypothesis. A parasitic reference must presumably be ineliminable, since every hypothesis trivially has logically equivalent reformulations that refer to other hypotheses. For instance, any theory T can be described in terms of any other theory T' via the formula (T & T') ∨ (T & –T').

The notion of parasitism is, obviously, tied to a syntactic account of theories. If, as writers like van Fraassen and Giere maintain, a theory is to be identified with a set of models, then it's impossible to draw the distinction between T and T' that Leplin and Laudan wish to make. I don't plan to make an issue of their syntacticism, however. My first criticism is that Leplin and Laudan provide us with no reason for supposing that T' is in fact parasitic. To be sure, the initial construction of T' makes reference to T. But this doesn't imply that T' can't be given an alternative characterization that circumvents the reference to T. Leplin and Laudan don't address such questions as whether T' is finitely axiomatizable. Certainly it's possible to devise very simple theories whose empirical consequences can be independently described. Leplin and Laudan's position seems to be based on the dual faith that one can't perform this trick every time (which is a nonobvious logical hypothesis) and that the distinction between the cases where one can and the cases where one can't perform the trick will correspond to their intuitions about charm (which I see no reason to expect).

To be sure, I haven't shown that T' is in every instance *non*parasitic. But this doesn't need to be shown to defuse the parasitism argument. Let's grant that the parasitism criterion succeeds in eliminating T' from the ranks of genuine rivals. In that case, I would argue that it's far too strong a test for the possession of charm, for there are circumstances where structures like T' have an important role to play in the game of science. Consider the following scenario: (1) theory T has been well confirmed, such that its empirical adequacy is widely believed; (2) it's discovered that one of its theoretical principles is inconsistent with an even more firmly believed theory; and (3) no one can think of any way to describe the empirical consequences of T—except as the empirical consequences of T. In that case, we might very well come to believe a proposition that has precisely the structure of T': that the empirical consequences of T are true, but that T itself is false. A recent example of this situation is the instrumentalist view of intentional psychology espoused by Dennett (1971) and accepted by a large portion of the cognitive science community. Dennett admits that intentional psychology has enormous predictive success—so enormous, in fact, that it would be foolish not to avail ourselves of its resources. But he refuses to accept its ontology on the grounds that it conflicts with physicalism. Furthermore, neither he nor anyone else knows how to get the same empirical consequences out of any other plausible theory. It's irrelevant here whether Dennett's position is correct. The important point is that his position is coherent and widely accepted by cognitive scientists. In a case like this, we can say nothing more than that the empirical consequences of T are true but that T itself is false. Surely we must allow this much into scientific discourse.

Now it's true that Dennett believes that there *exists* an independent physicalist theory with more or less the same empirical consequences as intentional psychology. Perhaps Leplin and Laudan are willing to consider such an existential claim, if it is true, as sufficient for passing the parasitism test: even if we don't yet know how to eliminate one theory's parasitic reference to another theory, the fact that we know it to be eliminable in principle is enough for us to allow that the first theory is a genuine rival. This approach makes our lack of deductive certainty about the eliminability of apparently parasitic formulations all the more salient. Given this uncertainty, the most that can be said of a putative quasi-hypothesis T' is that we presently *think* it is a quasi-hypothesis. We have to admit, however, that someone may find an independent characterization of T' in the future, in which case it will have to be promoted to the rank of a genuine rival. But surely we must agree to the following principle: if there's some chance that we will have to take a claim seriously in the future, then we already have to take it seriously now, albeit perhaps not *as* seriously. You can't summarily dismiss a hypothesis that may turn out to be charming. But the *raison d'être* for the notion of charm is to allow for summary dismissals.

There's another move available to proponents of a parasitism-based rejection of T'. Consider again the scenario in which we believe that the empirical consequences of T are true (call these consequences T^*) but that T itself is false. We might wish to claim, on the basis of a general argument, that if the empirical consequences of a theory are true, then there must be *some* true, unified, and independent theory that has just those empirical consequences. If this principle were true, then we could be sure that we'll never be in a position where we have to make scientific use of an ineliminably parasitic formulation. One problem with this line, of course, is that there seems to be no compelling reason to believe that every true T^* can be derived from a true unified theory. Even if it's agreed that every observable event in the universe can be obtained as a consequence of a single, unified, and elegant theory, the T^* in our possession may very well be a gerrymandered *piece* of the whole truth that fails to be identical to the consequences of any true and unified theory. To be sure, the true total theory of the universe will *include* T^* in its more vast set of empirical consequences. But if this kind of inclusion is sufficient to pass the parasitism test, then the test is vacuous.

Even more seriously, the foregoing line of defense undercuts the foundation of the parasitist's own position. To accommodate the fact that some apparently parasitic formulations are accepted as genuine rivals, it's admitted that the empirical consequences of one hypothesis *may* alternatively be described as the empirical consequences of an entirely different hypothesis. But this is to admit that there's no guarantee that T', the hypothesis that T^* is true but that T is false, is parasitic on T. In fact, if it's assumed that every true set of empirical consequences can be obtained as the empirical consequences of a true theory, then T' automatically entails its own nonparasitism on T.[1] And without this very strong assumption, the Dennett scenario requires us to admit either that parasitic formulations may be genuine rivals or that we have as yet no basis for believing of any T' that it is parasitic on T.

The final and, I think, most persuasive point against the parasitism argument is that there are alternative algorithms for producing empirical equivalents that are demonstrably nonparasitic. For any theory T, construct the theory T! which asserts that T holds whenever somebody is observing something, but that when there is no observation going on, the universe behaves in accordance with some other theory T_2 that is incompatible with T. More precisely, during intervals devoid of observational acts, events take place in accordance with theory T2, but when observation begins again, events once again proceed in a manner that is consistent with the hypothesis that theory T had been true all along. By definition, there is no possibility of finding traces of T2's rule among initial conditions when the rule of T is reasserted. The rival T! is empirically equivalent to T.[2] Moreover, the reference to T in T! is clearly eliminable. Whatever detailed story T tells about the world, we can describe T! by telling the same story and adding the proviso that this story is true only when the world is being observed. Thus, Leplin and Laudan can't hope to eliminate all the rivals to T by means of the parasitism criterion alone. The most that can be claimed is that nonparasitism is a necessary appurtenance of charming hypotheses. If we wish to eliminate T! as well as T', we have to add further requirements for charm.[3]

5.5 The Superfluity Criterion

Leplin and Laudan discuss and repudiate T! as well as T' in their 1993 essay. When they deal with T!, however, they rely—as they must—on a different criterion for charm. This argumentative strategy must immediately give us pause: if every proposed algorithm for producing empirically equivalent rivals is going to require us to come up with a new ad hoc rule to ensure its disqualification, the charm-based argument against EE loses all credibility. Leplin and Laudan claim that T! fails to be a genuine rival because there's no possible observation that would confirm its "additional commitment to a mysterious intermittency of natural law" (1993, 13). Evidently, a genuine hypothesis not only must be confirmable, but also must not contain any subset of principles that can be excised without loss of empirical consequences. Let's call this the *superfluity* criterion.

One problem with this charge against T!, which Leplin and Laudan themselves immediately note, is that the original theory T has an equally superfluous commitment to the hypothesis that its laws *continue* to hold when nobody is looking. Certainly T can't be produced simply by the deletion of superfluous portions of T!.[4] So why the preferential treatment for T? According to Leplin and Laudan:

> The reason is that there is overwhelming support, albeit of an indirect kind, for the indifference of natural law to acts of observation and, more generally, for the kind of uniformity that [T!] violates. (1993, 13)

I doubt that this is a fair assessment of the quantum-mechanical worldview of contemporary science. But this issue is in any case irrelevant to the topic at hand. Leplin and Laudan are supposed to be providing us with a criterion for *charm*, which is inherently a virtue *other* than those having to do with empirical support, whether direct or indirect. The fact that one description of the world is better supported by the evidence than another doesn't make the second a quasi-hypothesis—it makes it a *worse* hypothesis. It seems that Leplin and Laudan's discussion has the following structure: they dismiss T! on the grounds that it's a logico-semantic trick, and when asked for justification of this claim, they cite empirical evidence that favors T over T!. But if their empirical argument is accepted, the appeal to logico-semantic trickery isn't needed to get rid of T!. If Leplin and Laudan are right about the empirical inferiority of T!, then T! is simply not a good test case for their criterion for charm.

By the way, there *is* a hypothesis that is empirically equivalent to T but that can be produced by the deletion of empirically superfluous portions of T — namely, the hypothesis T* which asserts only that the empirical consequences of T are true and says nothing about T's theoretical entities. Thus, if taken literally, the superfluity criterion tells us that only T* is a bona fide hypothesis and that T should be dismissed as a quasi-hypothesis. This is very close to what antirealists have been trying to persuade us of all along. Of course, not even avowed antirealists have been so brash as to deny that T is a *hypothesis*. It's odd to suggest that putative theories that are logically stronger than necessary should on that account be regarded as quasi-hypotheses. The attempt to discover a weaker hypothesis that has the same scope as an existing hypothesis is a standard scientific endeavor. But nobody takes the success of such an endeavor to mean that the superseded hypothesis failed to be a genuine scientific hypothesis after all — it just means that the old hypothesis is no longer the *best* hypothesis available. Once again, if the basis for rejecting an algorithmic rival is that it isn't as good a hypothesis as the original, then its discussion doesn't belong in this part of the chapter — or in the corresponding part of Leplin and Laudan's article. This very different attack on EE is dealt with in Section 5.8.

5.6 The Criterion of Scientific Disregard

A passing remark of Leplin and Laudan's serves to introduce the next criterion:

> [Kukla] challenges us to explain wherein theoreticity [my old word for charm] exceeds logico-semantic trickery. Again, we counsel deference to scientific judgment as to what constitutes a theory. But as our case requires only the rudiments of an analysis, we accept the challenge. (1993, 13)

The part I wish to focus on is the recommendation that we defer to scientific judgment on matters of charm. Evidently, had they not been badgered into providing an analysis, Leplin and Laudan would have been content to let scientists tell them what to accept as a genuine theory without inquiring into the basis for this judgment. Is this a defensible attitude? We begin with the observation that scientists routinely and uniformly ignore hypotheses that seem to have a good measure of the traditional scientific virtues. Let's call this the phenomenon of *scientific disregard*. The deference to scientific disregard recommended by Leplin and Laudan might be rationalized in at least three ways. The most straightforward possibility is that scientists happen to be in possession of the right story about what makes hypotheses charming. Obviously, this scenario introduces nothing new into the discussion. In fact, all further discussion of the topic would be premature until somebody finds out from the scientists what the right story is. A second possibility is that scientists can no more articulate the criterion for charming hypotheses than we philosophers can — they make their judgments intuitively; however, the success of science gives us inductive grounds for believing that their intuitions are sound. This possibility is discussed in section 5.7. The third source of deference is also the most reverential of science. It may be supposed that scientific judgments about the charm of hypotheses are *constitutive* of their charm—that T! fails to be a genuine rival *because* scientists disregard it. On this view, the mere fact that a hypothesis is disregarded by science is reason enough, without further analysis, to brand it as a quasi hypothesis. Naturally, this course recommends itself to philosophers who regard the practices of science to be generally constitutive of rationality.

Note that even if scientific disregard were constitutive of the absence of charm, it would still make sense to inquire into how scientists make their determination. But the issue would cease to matter in the realism-antirealism debate. For if the grounds for eliminating putative rivals is the fact that they aren't taken seriously by scientists, then we can resolve the underdetermination issue without knowing exactly how scientists decide what to take seriously and what to disregard. All that needs to be ascertained is whether scientific theories universally confront empirically equivalent rivals that scientists actually worry about. And the answer to this historical question is unambiguously negative.

Has anyone actually used this argument against underdetermination from the principle of deference to scientific judgment? Perhaps Leplin and Laudan would endorse it. At any rate, I find it clearly *presupposed* in the following passage by Ronald Giere:

> Van Fraassen's view is that we must choose one hypothesis out of a literally countless set of possible hypotheses. This looks to be a very difficult task given finite evidence and the logical underdetermination of our models by even their complete empirical substructures. I would urge a quite different picture. . . . Looking at the scientific process, we see scientists struggling to come up with even one theoretical model that might account for phenomena which previous research has brought to light. In some cases, we may find two or three different types of models developed by rival groups. These few models may be imagined to be a selection from a vast number of logical possibilities, but the actual choices faced by scientists involve only a very few candidates. (1985a, 86–87)

On the face of it, what Giere is claiming is that, though there may exist numerous empirically equivalent formulations to any theory, scientists are generally unable to *think* of more than one or a few. On this reading, Giere's point is that the problem of underdetermination is resolved in practice by the limited abductive capacities of scientists. This idea that it's very difficult to come up with empirically equivalent rivals is also sometimes expressed by artificial intelligence researchers working on computer models of scientific discovery (e.g., Langley et al. 1987, 16–17). But what about T' and T!? It isn't plausible to suppose that scientists are so unimaginative that they're unable to conceive of these possibilities. Evidently, the claim that it's difficult to come up with empirically equivalent rivals is based on the presupposition that T', T!, and other peculiar constructions *don't count*. So, despite what Giere may appear to be saying, it isn't the case that the problem of underdetermination is resolved in practice by scientists' inability to think of empirically equivalent formulations. The problem is solved rather by scientists' unwillingness to take many easily conceivable formulations seriously. But to say that their mere unwillingness to take them seriously solves the problem is to endorse the criterion of scientific disregard.

What are we to say about this argument? Giere is well known for his antinormative perspective on the philosophy of science. I'm not so willing as he is—or as Leplin and Laudan seem to be—simply to defer to scientific judgment. If it turns out that scientists uniformly ignore certain hypotheses on the basis of their form, I would like to know what their thinking is behind the practice, and I reserve the option of deciding that this particular practice is nothing more than a historical prejudice that hasn't contributed to the success of science. But this is not the place for disquisitions into the role of normative versus descriptive analyses in the philosophy of science. The following criticism of the argument from scientific disregard is entirely compatible with Giere's extreme descriptivism.

Let us, for the sake of the argument, accept that scientific judgments are constitutive of rationality. There are two reasons why this admission doesn't invalidate all the algorithmic rivals of T. The first is that the algorithmic rivals have *not* been universally ignored by scientists. In section 5.4, I offered the counterexample of cognitive scientists' acceptance of T', where T is intentional psychology. In section 5.7, I discuss a different algorithmic rival that is taken seriously by at least some contemporary physical scientists. Second, even if all the algorithmic rivals were universally ignored by scientists, it wouldn't follow that these rivals suffer from any epistemic failings. What is uncontroversial about scientific practice is that scientists never *pursue* more than a few theories at a time. This observation doesn't yet establish that scientists consider these few theories to be *epistemically* superior to their ignored empirically equivalent rivals. Van Fraassen (1980) claims that only a theory's empirical content counts in its epistemic evaluation; therefore, the decision to ignore the algorithmic empirical equivalents to our theories must be merely "pragmatic". Of course, van Fraassen's assumption about the epistemic irrelevance of nonempirical virtues may be mistaken. But regardless of whether we agree with this assumption, realists and antirealists alike must concede that there *are* nonepistemic theoretical virtues that influence theory pursuit. For example, we might pursue a theory because it's easier to teach to undergraduates than its rivals, or because it's favored by the granting agencies. No one would suppose that such virtues should count in the determination of which theory to believe. Therefore, it's possible that what distinguishes the "live" theories of science from their many disregarded empirical equivalents is a nonepistemic virtue. If this is so, then the problem of underdetermination—the problem that we have no more reason to believe any theory than any of its many empirical equivalents—remains untouched by the phenomenon of scientific disregard. This counterargument doesn't refute a position like Giere's, but it shows it to be in need of further support. What's needed is a demonstration that the empirically virtuous but disregarded hypotheses of science are ignored for epistemic reasons.

5.7 The Intuitive Criterion

As far as I know, parasitism, superfluity, and scientific disregard are the only candidate criteria for charm to be found in the philosophical literature. If the conclusions of the last three sections are accepted, we are thus left without a workable criterion on the table. This state of affairs does not, by itself, compel us to abandon the notion

of a quasi-hypothesis. If the lack of an explicit criterion for its application were reason enough to banish a philosophical concept, there would be very few concepts for
philosophers to write about. Defenders of a charm-based refutation of EE could plausibly adopt the following line: it may be true that we're unable to specify precisely
what properties T' and T! possess that relegate them to the realm of quasi-hypotheses,
but neither are we able to cite the precise criterion whereby Newtonian mechanics
is judged to be a theory of great explanatory scope. In both cases, however, the truth
of the claim is intuitively obvious. To be sure, judgments of intuitive obviousness
have sometimes been overturned, but so has every other kind of judgment. I concede, at least for the sake of the argument, that all other things being equal, intuitive
obviousness is an acceptable ground for belief. Moreover, I think I can access the
same store of intuitions about the charm of hypotheses as HR and LL. The idea that
T' or T! might be the whole of the truth strikes them and me as *unintelligible* in some
sense. Perhaps it's the lack of a similar sort of intelligibility that makes quantum
mechanics a perennial target for antirealist interpretation. I do not pursue the analysis of intelligibility here. I proceed under the assumption — or rather the hope — that
our intuitions in this area agree. My claim is that there are algorithms for producing
empirically equivalent rivals to T that *are* intuitively intelligible — at least to me and
to some physical scientists.

The algorithm I have in mind is based on a speculative hypothesis discussed by
the astronomer John Barrow (1991). Barrow presents it in the course of evaluating a
hypothesis that is far removed from our present concerns. It isn't clear from the text
whether he grants his construction any appreciable degree of plausibility. But it *is*
clear that he regards it as a genuine hypothesis rather than a quasi-hypothesis. The
basic idea is that for any theory T about the universe, there is another theory A(T)
which asserts that what we call the universe is a computer simulation wherein events
are programmed to follow the rules of T:

> If we were to build a computer simulation of the evolution of a small part of the
> Universe, including, say, a planet like the earth, then this model could in prin-
> ciple be refined to such an extent that it would include the evolution of sentient
> beings who would be self-aware. They would know of and communicate with other
> similar beings which arise within the simulation and could deduce the program-
> ming rules which they would designate as "laws of Nature". . . . Indeed, we could
> be the components of such a simulation. (185)

I want to claim that T and A(T) are empirically equivalent. This is not as entirely
obvious as I would like it to be. We might reasonably wonder, as we do about the
kindred hypothesis that we are brains in a vat, whether the reference of the observational terms used in T is the same as the reference of the corresponding terms in the
alternative hypothesis. This locus of uncertainty can be finessed by a judicious change
in Barrow's scenario. Let A(T) stipulate that a race of beings exists — call them the
Makers — who create or assemble a universe in which the theoretical and observational entities of T literally exist. In this case, T and A(T) are not only empirically
equivalent — they even agree on many theoretical details.

In fact, as far as we have taken the description of A(T), it agrees too much with
T: if *all* the consequences of T, both empirical and theoretical, are consequences of

A(T), then, even if A(T) is a genuine hypothesis, it still fails to be a genuine *rival* to T. A(T) just has more to say than T. But this problem is easily fixed. Consider Barrow's original computer simulation version of A(T). Aside from problems having to do with identity of reference, there's an obvious source of potential disagreement between T and A(T). When the programmers who are responsible for the simulation *turn the computer off*, our world ceases to exist, and when they turn the computer on again (having stored the closing configuration in memory), our world pops back into existence. Naturally, we simulated beings would have no knowledge of this intermittency in our existence. Our theory T thus has at least this disagreement with Barrow's A(T): T asserts that certain theoretical entities exist continuously, while A(T) asserts that they exist intermittently. The same feature can be adapted to the scenario of the Makers. Let's say that the continued existence of the worlds created by the Makers depends on the continuous operation of a machine, and that this machine is occasionally turned off for maintenance and repair. This A(T) is an algorithm for producing hypotheses that are both empirically equivalent to T and incompatible with T. Moreover, both Barrow and I deem A(T) to be intelligible.[5] By the intuitive criterion, A(T) is thus a genuine rival to T. In fact, except for a few cosmetic changes, A(T) tells the same story as T! but adds to it an account that renders the curious intermittencies of T! intelligible. A(T) provides us with a picture of how intermittencies like those postulated by T! might come to be.

A similar algorithm—call it B(T)—provides us with an intelligible version of T'. In B(T), as in A(T), our observations are determined by the activities of another race of beings—call them the Manipulators. Unlike the Makers, the Manipulators don't have the technology for creating worlds in which the theoretical entities of T exist. They live in the same world as we do. However, they manipulate events in such a way that all our observations are consistent with the hypothesis that T is true (perhaps they're psychologists who wish to test our ability to abduce T when given the appropriate evidence). It may be supposed that the Manipulators use various ad hoc devices, not excluding sleight of hand, to achieve their aim of sustaining the appearance of T. However they accomplish their mission, the net result is that we find T to be empirically adequate. B(T), too, is an intelligible empirically equivalent rival to T.

B(T) is a minor variant of the Cartesian story about the evil genius. Descartes was also concerned with providing experientially equivalent rivals to certain views (although in his case, the target view was not a scientific theory, but commonsense realism about the objects of perception). As far as mere experiential equivalence goes, he could simply have cited the alternative that corresponds to T', namely, that our perceptions have just the character that they would have if they were caused by physical objects, but that they are not so caused. It was presumably his desire to provide an *intelligible* empirical equivalent—a "genuine rival"—that led to his adding the bit about the evil genius.

It might be objected against both A(T) and B(T) that it isn't obvious that they describe logically coherent possibilities. I concede the point: it's possible that further analysis will show that one or both of these stories is unrealizable on logical grounds. But is *anything* obviously coherent? After all, it's happened more than once that theories in good conceptual standing have ultimately been rejected on the grounds of incoherence. Look at what Russell did to Frege. Deductive infallibility is just as elusive a goal as its inductive counterpart. It would be unreasonable to demand that every new concept or principle be introduced together with a demonstration of its coherence (whatever such a demonstration might be like). It's a universal and, I think, pragmatically unavoidable practice to *presuppose* the coherence of an idea until arguments to the contrary arise.[6] The burden of the argument is thus on those who wish to claim that A(T) and B(T) are incoherent in a way that can't be fixed by straightforward changes in their description. Speaking of logical coherence, it should be recalled that deniers of EE have been allowed to get to this point in the argument only by grace of the generous assumption that the concept of charm itself is coherent. It seems at least as likely that charm makes no sense as that A(T) and B(T) make no sense.

Finally, note that Barrow's discussion of A(T) produces a dilemma for those who wish to rely on the criterion of scientific disregard to refute EE. Since Barrow and other scientists accept A(T) as a genuine rival to T, deniers of EE must either admit that they have lost the argument or retract their counsel of "deference to scientific judgment as to what constitutes a theory" (Leplin & Laudan 1993, 13).

5.8 On the Putative Epistemic Inferiority of Algorithmic Rivals

A different sort of refutation of the underdetermination argument has been presented by McMichael (1985). McMichael tacitly concedes that the algorithmic construction T' is empirically equivalent to T, and that it's a genuine rival to T. But, he claims, although T' is in the competition for best hypothesis, there's a general argument which shows that T' must always and inevitably lose to T. According to McMichael, we're justified in systematically preferring T over its rival T' on the grounds of the former's greater simplicity:

> [T'] will always lack the simplifying power of the theory it is intended to replace, since it postulates each observable regularity as a fundamental law rather than postulating, more plausibly, that this complex collection of regularities is a consequence of a relatively small set of primitive theoretical truths. (270–271)

Evidently, this argument requires us to deny the second antirealist premise that EE entails UD.[7] This aspect of McMichael's argument takes us beyond the terms of reference of the present chapter. In this section, I grant the general principle that nonempirical virtues like simplicity may have epistemic import, and evaluate McMichael's particular version of this thesis. I also say a few things at the end about the general argumentative strategy of trying to show that the algorithmic rivals are in the running but that they always lose.

Before evaluating McMichael's proposal, let's take a few moments to differentiate it more sharply from Leplin and Laudan's proposal about charm. Couldn't we equate the notion of quasi -hypothesishood with having an extreme value of zero on a scale of simplicity, and thereby reduce Laudan and Leplin's solution to a variant of McMichael's? The problem with this suggestion is that the notion of an absolute minimum of simplicity is incoherent. Perhaps it's plausible to talk about *maximal* simplicity, but it's constitutive of the concept of simplicity that, given any hypothesis, we can always construct a *less* simple one from it by injudicious accretions. What about imposing a threshold of simplicity below which hypotheses will be considered quasi-hypotheses and eliminated from further consideration, regardless of how virtuous they may be in other respects? This is certainly not McMichael's theory. McMichael doesn't claim that T' falls below a threshold of required simplicity. His argument is rather that, whatever the level of simplicity may be for T', we are assured that the empirically equivalent T is going to be *simpler*; hence, T is always to be preferred. Also, the simplicity-threshold account of charm has quite unacceptable consequences. Suppose that a hypothesis X is slightly below the threshold for simplicity while Y is slightly above threshold, but that X has enormously more empirical support than Y. Surely we can make the difference in simplicity so small and the difference in empirical support so large that it would be folly to take Y seriously as a genuine scientific competitor but to dismiss X as not even being in the game. No hypothesis can be regarded as *absolutely* worthless on the basis of its position on a continuous dimension of simplicity. There are only more or less simple hypotheses. Therefore, it's possible for the other virtues of a hypothesis to be so much greater than its competitors' that we are willing to overlook its failings in the simplicity department. McMichael's view is that this systematically fails to happen in the case of the algorithmic rivals. In contrast, Laudan and Leplin's concept of charm refers to a *discrete* property of propositions that rules potential competitors out of bounds before any other evaluation is made. Whatever other virtues a quasi-hypothesis may have, they will never be sufficient to get it into the game of science. For Laudan and Leplin, the algorithmic rivals are just not the kind of thing that scientists are looking for; for McMichael, they're the right kind of thing, but they can be shown always to be inferior to the theories from which they're derived.

My criticism of McMichael's simplicity criterion doesn't contain any philosophical news. It's notorious that most criteria of simplicity produce results that are dependent on a prior choice of language. The idea that T covers the phenomena with a few laws while T' requires many laws to do the same job suffers from this problem. On the one hand, any one of the laws of T can be rewritten as an indefinitely numerous set of sublaws. "F = ma", for instance, is logically equivalent to the conjunction of "F = ma in New York", "F = ma in Philadelphia", and "F = ma everywhere outside of New York or Philadelphia". On the other hand, any number of laws can be rewritten as a single statement asserting their conjunction. Nor is this second problem resolvable by a simple prohibition against conjunctive laws. For example, consider this famous pair of observable regularities:

(1) All emeralds are green.

(2) All sapphires are blue.

Define an *emire* as anything that is either an emerald or a sapphire, and a *gremblire* as anything that is either a green emerald or a blue sapphire. Then the conjunction of (1) and (2) is logically equivalent to the following single, conjunctionless rule:

(3) All emires are gremblires.[8]

We can't blithely assume that there is such a thing as the absolute number of laws in a system. Those who wish to rely on this notion are obliged to provide a method of enumeration. As far as I know, the only serious attempt to tackle this problem has been by

Michael Friedman (1974), and Salmon (1989) has shown that even Friedman's carefully crafted proposal fails in the end.

McMichael has some things to say about the greenness of emeralds. He notes that van Fraassen feels justified in believing "All emeralds are green" over the empirically *non*equivalent "All emeralds are grue", and argues that this preference commits him to accepting that some nonempirical virtues like simplicity are relevant to what we believe. This is a criticism of the second premise of the underdetermination argument; hence it is, once again, beyond the purview of the present chapter. Even if this criticism is accepted, however, it doesn't show that it's *simplicity*, as McMichael conceives it, that provides the solution to the grue problem. The criticism of his notion of simplicity stands on its own ground. If it's true that McMichael's simplicity is language dependent, then either we must have a *different* nonempirical reason for preferring "All emeralds are green" to "All emeralds are grue", or we must concede that our epistemic preference depends on our choice of a language and we must all become relativists. Whichever of these two routes is taken, McMichael's argument for believing T rather than T' will not go through.

But what if the choice is between an indeterminate finite number and infinity? If the empirical content of a theory is *not* finitely axiomatizable, then it might not matter that we can't say precisely how many principles there are in the theory. The fact (if it is a fact) that T is finitely axiomatizable while its empirical content isn't would be enough to establish the greater simplicity of the former. McMichael doesn't explore the issue of finite axiomatizability. It used to be thought that the issue was settled in the affirmative by Craig's theorem, according to which the set of all consequences of T that use observational predicates is finitely axiomatizable. But van Fraassen has reminded us that the set of observation-language consequences of a theory is not to be equated with the theory's empirical content (1980, 54–55). Still, McMichael provides us with no reason to suppose that the empirical content of T *can't* be finitely axiomatized.[9]

The simplicity criterion is even more clearly unable to block T! from winning the competition against T. In the case of T!, there's no question of pitting an infinite set of principles against a finite set: T! is finitely axiomatizable if T is. So the claim that T wins over T! because of its greater simplicity must be based on an actual enumeration of principles. Does T have fewer principles than T!? To be sure, we've expressed T! as a conjunction — roughly, T! states that the laws of T hold when the universe is observed and that the laws of T2 hold when the universe is not observed. But this can't be a reason for considering T! to be more complex than T, for T is logically equivalent to the syntactically parallel statement that the laws of T hold when the universe is observed and that the laws of T also hold when the universe is not observed. It's been suggested to me that T might be simpler than T! by virtue of the fact that, even when they are both formulated in parallel conjunctive form, T mentions only one theory whereas T! mentions two theories. This defense of McMichael's claim begs the question of whether there's an objective manner of enumerating principles. It's true that the conjunctive form of T! mentions different theories (T and T2) in each of its conjuncts, while the conjunctive form of T mentions the same theory in each conjunct. But there are other ways of representing the claims of T and T! in which this relation is reversed. For example, T may be represented as the conjunction "T! is true when the world is observed, and T is true when the world is not observed"; and T! can be represented by the simpler conjunction "T! is true when the world is observed, and T! is true when the world is not observed."

The repudiation of McMichael's criterion does not, of course, preclude the possibility that the algorithmic rivals might lose the competition on the basis of some other nonempirical virtue. Regardless of which virtue is alluded to, however, this theoretical strategy is not going to be able to accomplish much even if it's successful. Suppose it were established that all the algorithmic rivals have a lesser degree of some nonempirical virtue. Such a result would count against the argument from underdetermination only if it were also established that nonempirical theoretical properties may have epistemic import. But if *that* were to be accepted, then the second premise of the underdetermination argument—the claim that EE entails UD—would stand refuted, and the underdetermination argument against realism would *already* have failed. There would be no need to establish that the algorithmic rivals always lose. For the purpose of protecting scientific realism from the underdetermination argument, it doesn't matter whether the algorithmic rivals win or lose. It matters only that something wins.

5.9 The Denouement

I've considered several strategies for discrediting several algorithms that produce empirically equivalent rivals to any theory. None of these proposals succeeds. Given the current state of the debate, the rational choice is to accept EE as true. This means that the status of the underdetermination argument depends entirely on whether its second premise is also true.

It's time to return to the point, made in section 5.1, that there are several varieties of EE, each of which (in conjunction with the second premise) entails the falsehood of a different version of scientific realism. Which variety of EE do I take to have been established? If all my arguments are accepted, then it's established that every theory has at least some rivals that do not posit theoretical entities (e.g., T'), as well as some rivals that do posit theoretical entities (e.g., T!). This result is sufficient to refute both concrete realism (rationally warranted belief in some specific set of theoretical entities) and abstract realism (rationally warranted belief in the proposition that there are theoretical entities)—always assuming, of course, that the second premise of the argument from underdetermination is true. But it still leaves both forms of feeble realism in the running—the rational warrantability of ascribing subdoxastic but nonzero probabilities to (a) some hypothesis of the form "theoretical entity X exists" (feeble concrete realism) or to (b) the hypothesis that there are theoretical entities (feeble abstract realism).

But if we grant that T' is a rival to T, then we can extend the number of theoretical rivals indefinitely by constructing theories of the form T' & U_i, where the U_i run through an infinite sequence of theoretical hypotheses that have no empirical consequences, either alone or in conjunction with any other hypothesis. For example, U_i could be the hypothesis that there are i families of elementary particles that enter into no interactions with any other particles in the universe. If we accept that this algorithm produces infinitely many rivals that posit different theoretical entities, then even feeble concrete realism is refuted. And why should the algorithm not be accepted? To complain about the implausibility of the theories it produces is illegitimately to presuppose that the second premise of the argument from underdetermination is false (for otherwise, two empirically equivalent theories cannot possess different degrees of plausibility). To complain that T' & Ui isn't even in the running is to precipitate a recapitulation of the foregoing arguments about charm. It might be argued that all theories of the form T' & Ui are logically stronger than T', hence that they lose the epistemic competition to T' on the grounds of probability theory alone. Now, T' is itself an algorithmic construction from T, so this championing of T' would only be proposed by realists engaged in a desperate attempt to hold the line at the severely retrenched position of feeble concrete realism. The problem with this argument, however, is that the same principle leads to the conclusion that both T' and T lose to T*, the hypothesis that says that T is empirically adequate, but that doesn't say anything about the truth or falsehood of T. That is it say, it's an argument for antirealism.

That leaves us with feeble abstract realism. To expel this last toehold of realism, it would have to be shown that there are infinitely many incompatible rivals to T that do not posit theoretical entities. I concede that there is as yet no argument that establishes so strong a form of EE. Let there be rejoicing in the camp of the realists.

6

Realism and Underdetermination II

Does Empirical Equivalence
Entail Underdetermination?

Now for the second premise. In this chapter, I deal with arguments that try to refute the claim that empirically equivalent theories must also be epistemically equivalent. Considerations in support of the second premise are taken up in chapter 7, where they appear as part of a more general defense of antirealist themes.

6.1 The Nonempirical Virtues of Theories

Almost all the arguments against the second premise are based on a single idea: that there are global properties of hypotheses, such as their simplicity or explanatoriness, that are wholly independent of the hypotheses' empirical content but that nevertheless can have an effect on their epistemic status. Let's call the principle that such arguments *deny*—namely, that there are *no* nonempirical virtues that have epistemic import—by the name of NN. The presumed connection between NN and the second premise is clear: if some nonempirical virtues do have epistemic import, then empirically equivalent theories aren't necessarily epistemically equivalent.

Realists have argued against NN in three ways. The first type of counterargument makes the straightforward claim that some specific nonempirical virtue provides a reason for belief. Glymour (1984), for example, maintains that a theory's capacity to explain phenomena is a reason for believing it. To my knowledge, no claims of this type have ever been based on a more fundamental principle that antirealists might be willing to accept. The discussions they engender therefore quickly degenerate into profitless clashes of intuitions.

The second type includes arguments to the effect that antirealists themselves, unless they accede to a thoroughgoing skepticism, must be willing to accept some specific nonempirical property as a reason for belief. Some writers have charged that van Fraassen's willingness to believe that all emeralds are green rather than grue commits him to the view that simplicity has epistemic import (e.g., Musgrave 1985). According to this argument, antirealists stand accused of internal inconsistency:

they're evidently willing to deploy certain nonempirical criteria in deciding which *empirical* hypotheses to believe among the many nonequivalent sets of empirical hypotheses that include the data, but they refuse to apply the same criteria for breaking ties among empirically equivalent *theoretical* hypotheses. This second argumentative strategy has a better chance of succeeding than the first, since it doesn't require us to justify the epistemic significance of the nonempirical feature in question. We need only show that antirealists need it to make the sorts of epistemic choices that they themselves regard as unobjectionable.

The third strategy is even more promising than the second. If, as in arguments of the second kind, we identify the precise nonempirical feature that presumably underlies theory choice, it's open for the antirealist to deny that *this particular feature* does in fact play the role ascribed to it. This is how van Fraassen responds to the claim that he must appeal to simplicity to believe that all emeralds are green — he simply denies it and rehearses all the ways of getting into trouble with the hypothesis that simplicity is epistemically significant (van Fraassen 1985). But the accusation of inconsistency can be made without pointing a finger at any particular nonempirical feature of hypotheses. Realists need only argue that the epistemic choices sanctioned by antirealists require them to use *some* nonempirical criterion. If simplicity doesn't work, then it must be something else. Even if we're unable to specify what it is, the fact that antirealists must appeal to *something* nonempirical presumably shows that they can't consistently maintain that NN is true. This is how the inconsistency argument is made by Hausman (1982), McMichael (1985), and Clendinnen (1989). In what follows, I ignore the first two types of arguments and concentrate on the third.

Let's try to state the charge of internal inconsistency more carefully. According to NN, the empirical virtues of a hypothesis are the sole determinants of its epistemic status. Now, there are indefinitely many *empirical* hypotheses that account for any set of data that we might already possess but that make different empirical predictions. Our observations to date are equally consistent with the hypothesis that all emeralds are green, with the hypothesis that all emeralds are grue, and indeed with infinitely many other hypotheses about the color of emeralds. The hypothesis of (epistemic) choice is, ideally, the one all of whose predictions turn out to be correct (in the case of empirical hypotheses, there are no empirically equivalent rivals to worry about). But we don't *know* which hypothesis is going to enjoy this continued predictive success. In fact, we will *never* know it. At any stage in the evolution of science, the most we will ever know is whether or not a hypothesis is consistent with all the data that have been established to date — and there will always be infinitely many gruish rivals that meet this criterion but make different predictions about what happens next. Yet, on the antirealist's own account, we are sometimes justified in believing that some of these hypotheses are more likely to be predictively successful than others. Such a belief must be based on nonempirical considerations — for what empirical considerations are there left to bring into play? If our choice isn't plausibly made on the basis of relative simplicity, or explanatoriness, then the operative criterion must be something else. We don't need to know what it is to be certain that the antirealist is committed to *some* such criterion. But to admit the existence of a nonempirical dimension of epistemic evaluation is to deny NN.

Let's suppose that the foregoing argument is sound. Then NN can't consistently be maintained by anyone other than a thoroughgoing skeptic. This counterargument fails to undermine the second premise, however, because NN isn't a necessary corollary of the second premise. Granted that there must be a rule for selecting among empirical hypotheses, it doesn't follow that such a rule must also enable us to select among empirically equivalent theories. Clendinnen has claimed otherwise. In the following passage, (P1) is a gruish theory that is empirically equivalent to theory (S) and (P2) is a gruish theory that accounts for the same data as (S) but is empirically distinct from (S):

> If there is to be any rational way of predicting then there must be some way of ruling out (P2) as unprojectable, for some such hypothesis could be used to predict anything. I have suggested simplicity used comparatively; but if not this, then something else. Whatever it is, a criterion which discriminates against the needless complexity of (P2) will surely also discriminate against the needless complexity of (P1). (1989, 72–73)

If Clendinnen is right—if a rule that selects (S) over (P2) must "surely" also select (S) over (P1)—then the antirealist's position is internally inconsistent. But this "surely" cannot have the force of logical entailment. For any epistemic discrimination we may wish to make among empirically *distinct* hypotheses, there are rules that will permit us to make that discrimination but that will fail to discriminate among empirically equivalent theories. If nothing else, we could simply *stipulate* that the rule applies only to empirically distinct hypotheses, just as we stipulate that the rules for division don't apply to zero denominators.

This retort is the same as the one made in section 2.4 to Boyd's (1984) defense of the miracle argument. Boyd claimed that scientists routinely appeal to explanationist virtues in assessing the merits of first-level scientific hypotheses, hence that it must be permissible to appeal to explanationist virtues in assessing the merits of metahypotheses such as realism and antirealism. Clendennin now claims that antirealists routinely appeal to nonempirical virtues in breaking epistemic ties among nonequivalent empirical hypotheses, hence that they're committed to breaking ties among empirically equivalent theories on the same basis. My point in both cases is that appealing to explanationist or other nonempirical virtues to settle one sort of issue doesn't logically commit you to appealing to them in other sorts of issues. To be sure, such a maneuver lays one open to the charge of being *arbitrarily selective* in one's skepticism. Quite likely, this is all that Clendinnen wished to claim. At any rate, the inconsistency charge has more the nature of a question to antirealists than a refutation. The question is: Why should our rules for hypothesis selection be restricted to empirically distinct hypotheses? I try to tackle this issue in chapter 7. But first, there are other attacks on the second premise to be dealt with.

6.2 Indirect Confirmation

In section 5.2, I discussed two arguments of Laudan and Leplin's (1991) which try to show that EE, the first premise of the underdetermination argument, is false. This pair of arguments constituted the first part of a two-pronged attack on underdetermination. The second prong, to which I now turn, consists of a pair of arguments that are directed against the second premise. Laudan and Leplin claim that even if EE is true, there's no rational way to get from there to UD. The arguments are (1) that hypotheses are sometimes confirmed by empirical facts that are not their own empirical consequences, and (2) that hypotheses sometimes *fail* to be confirmed by empirical facts that *are* their own empirical consequences. Each of these assertions seems to provide us with a method of blocking the move from EE to UD. For suppose that T_1 and T_2 are empirically equivalent rivals. To say that they're empirically equivalent is to say that they have the same empirical consequences. But if it's true that some nonconsequences may also be confirming, then one of the empirically equivalent theories may receive additional confirmation from a nonconsequence that the other theory fails to obtain, thereby rendering the first theory more believable than its rival. The same conclusion follows from the thesis that some consequences are nonconfirming, for in this case one of the theories may *fail* to be confirmed by one of its consequences and thereby become *less* believable than its rival. Let's look at each of these arguments in turn.

Here's the scenario that is supposed to establish that nonconsequences may be confirming:

> Theoretical hypotheses H_1 and H_2 are empirically equivalent but conceptually distinct. H_1, but not H_2, is derivable from a more general theory T, which also entails another hypothesis H. An empirical consequence e of H is obtained. e supports H and thereby T. Thus e provides indirect evidential warrant for H_1, of which it is not a consequence, without affecting the credentials of H_2. Thus one of two empirically equivalent hypotheses or theories can be evidentially supported to the exclusion of the other by being incorporated into an independently supported, more general theory that does not support the other, although it does predict all the empirical consequences of the other. (Laudan 1996, 67)[1]

In a situation such as Laudan and Leplin describe, let's say that the datum e provides *indirect confirmation* for the *destination hypothesis* H_1, and that this indirect confirmation is delivered to H_1 via the *bridge hypothesis* T.

Unlike the arguments discussed in section 6.1, this attack on the second premise doesn't depend on our being able to establish the epistemic significance of a nonempirical virtue. What breaks the epistemic tie between hypotheses H_1 and H_2 is a bona fide empirical fact. Nevertheless, this tiebreaker is at least in principle available to empirically equivalent theories, for the empirical fact that does the job isn't an empirical consequence of either of the tied hypotheses. The case for the validity of indirect confirmation is incomparably stronger than the case for the epistemic significance of any of the nonempirical virtues. Laudan and Leplin describe indirect confirmation as "the intuitive, uncontroversial principle that evidential support flows 'downward' across the entailment relation" (Laudan 1996, 67). I agree that the principle is both intuitive and uncontroversial. In fact, I'm willing to go further. In at least some of its applications, the principle of indirect confirmation can be justified on the basis of what I called the "Friedman maneuver" in chapter 4: when T's empirical consequence e is obtained, T may receive so great a confirmatory boost that its probability comes to exceed the probability that we had ascribed to H_1, the destination hypothesis that also follows from T but that isn't implicated in the derivation of e. But since H1 is a consequence of T, it's incoherent to suppose that p(T) is greater than p(H1). To restore coherence, the value of p(H1) must be elevated at least to the level of p(T). Thus, the validity of indirect confirmation is assured by the requirement of probabilistic coherence. Some people may also wish to maintain that evidential support "flows downward" across the entailment relation even if this probabilistic adjustment isn't forced on us. This is a debatable point. But it isn't debatable that evidential support flows downward when the general hypothesis T receives so much confirmation that p(T) comes to exceed the probability of one of its consequences. In cases like this, the epistemic significance of indirect confirmation rests on the solid ground, shared by realist and antirealist alike, of the probability calculus. (For philosophers like Laudan and Friedman who are averse to the idea of theoretical probabilities, indirect confirmation can be said to be based on the equally solid thesis that no hypothesis can enjoy a greater degree of confirmation than any of its consequences.) The phenomenon of indirect confirmation establishes beyond reasonable doubt that the believability of a hypothesis isn't always fully determined by its own empirical content. In chapter 4, I claimed that Friedman failed in his attempt to convert the validity of indirect confirmation into an argument for scientific realism. Laudan and Leplin want to use indirect confirmation in the service of the more modest goal of refuting an argument for antirealism. It's prima facie possible that this endeavor can succeed despite the failure of Friedman's project.

But it doesn't succeed. The argument from indirect confirmation is vulnerable to the same criticism that I mounted against the "classical" refutation of premise 2 in section 6.1. According to the earlier argument, antirealists' preference for "All emeralds are green" over "All emeralds are grue" must be based on their attributing epistemic significance to some nonempirical property of hypotheses such as simplicity. But then (the classical argument goes), if it's admitted that there are epistemically significant nonempirical properties of hypotheses, these properties are available for breaking epistemic ties between empirically equivalent theories. My criticism of this argument was that the desired conclusion isn't necessitated by the concession that was wrung out of the antirealists: antirealists can admit that they need to appeal to nonempirical properties of hypotheses to justify their empirical beliefs without also admitting that they have to use these properties to break theoretical ties. The same response is also applicable to indirect confirmation. Here again, even if the putative source of epistemic significance is admitted, antirealists have the option of denying that this source is ever available for breaking ties between empirically equivalent theories. The difference between the two cases is that the epistemic significance of nonempirical properties of hypotheses is controversial, whereas the epistemic significance of indirect confirmation *has* to be granted. The rest of the story is the same, however.

The crucial misstep in Laudan and Leplin's analysis occurs when they stipulate that obtaining the empirical consequence e of theory T results in increased support for T. In the context of a defense of scientific realism against an antirealist attack, it simply isn't permissible to assume that theories are epistemically strengthened by the discovery that their empirical consequences are true. To suppose this is already to presuppose that antirealism is false. If defenders of realism were allowed to say that e confirms T, they wouldn't need to say anything more. The battle would already be over. Recall that Laudan and Leplin have assumed (for the sake of the argument) that EE is true. But if EE is true, then there are indefinitely many theories that have the same empirical consequences as T. To suppose that we can ascertain which one deserves the confirmatory boost from e is already to suppose that the problem of underdetermination has been resolved.

I want to emphasize that the foregoing critique completely allows for the rationality of indirect confirmation: *when* a bridge hypothesis receives sufficient *direct* confirmation, *then* the confirmatory effect "flows downward across the entailment relation". But this principle doesn't tell us when the antecedent condition is satisfied and when it isn't. Antirealists may very well agree that indirect confirmation is a valid mode of inference. Indeed, it's my view that they're required to agree to this. But they can also claim without inconsistency that indirect confirmation never helps to break theoretical ties, because when the bridge hypothesis is a *theoretical* hypothesis, it never receives a nonzero supply of confirmation to disburse in the first place. Antirealists are going to want to say this because they believe in the underdetermination argument. Of course, if the underdetermination argument doesn't work, they may not have a leg to stand on. But you can't assume that the underdetermination argument doesn't work in the course of a proof that it doesn't work!

What about turning the argument into a species of scientific practice argument? The claim would be that, as a matter of actual practice, when the consequences of a bridge theory like T are found to obtain, scientists become more confident of the predictions that follow from the destination theory H_1. If this is granted—and if it's granted that our philosophy of science has to respect the practice—then it might seem that we're required to admit that e confirms T. If we do admit that e confirms T, then Laudan and Leplin's proof that EE doesn't entail UD is able to get off the ground. Of course, it would no longer *need* to get off the ground, for as soon as it's admitted that e confirms T, the realist passengers who want the argument to fly would already have arrived at their destination. But it isn't true that the scientific practice in question requires us to admit that e confirms T. The practice can just as well be rationalized by the alternative antirealist assumption that e merely confirms T^*, the hypothesis which asserts only that the empirical consequences of T are true. For if T entails H_1, then any empirical consequence of H_1 is also an empirical consequence of T. That is to say, T^* entails H_1^*. But then, if e produces a big confirmatory boost in T^*, $p(T^*)$ may come to exceed $p(H_1^*)$, whereupon $p(H_1^*)$ will have to be elevated. The result will be just what the scientific practice requires: an increased confidence in the empirical consequences of the destination theory. Of course, any benefit that accrues to H_1^* is automatically also reaped by H_2^*, since $H_1^* = H_2^*$. On the other hand, T^* doesn't entail H_1 any more than it entails H_2. Thus, no matter how big a boost T^* gets, the mechanism of indirect confirmation doesn't affect the epistemic deadlock between H_1 and H_2. In sum, one can allow both (1) that indirect confirmation is a valid mode of inference and (2) that the occurrence of e in Laudan and Leplin's scenario ultimately results in our having more confidence in the empirical consequences of H_1, while (3) still denying that e confirms T. But then (1) and (2) aren't enough to show that EE doesn't entail UD.

6.3 Epistemically Insignificant Empirical Consequences

Laudan and Leplin also claim that some empirical consequences fail to confirm the theories from which they can be derived. Here's their main example:

> Suppose a televangelist recommends regular reading of scripture to induce puberty in young males. As evidence for his hypothesis (H) that such readings are efficacious, he cites a longitudinal study of 1,000 males in Lynchburg, Virginia, who from the age of seven years were forced to read scripture for nine years. Medical examinations after nine years definitely established that all the subjects were pubescent by age sixteen. The putatively evidential statements supplied by the examinations are positive instances of H. But no one other than a resident of Lynchburg, or the like-minded, is likely to grant that the results support H. (Laudan 1996, 68)

Why is this state of affairs supposed to be an embarrassment for antirealists? Presumably, because the move from EE to UD is based on the presupposition that all empirical consequences provide evidential support for the hypotheses from which they can be derived. If this were not so, then it would be possible for e to be an empirical consequence of each of two empirically equivalent theories T1 and T2, but for e *not* to support T1, the result being that T2 would enjoy more total support. There are a few knots in this line of reasoning that I try to unravel below. If the reasoning were to be accepted, however, Laudan and Leplin's hand could be substantially strengthened by noting that antirealists need an even stronger presupposition to get from EE to UD: they need to assume that every empirical consequence provides *the same amount* of evidential support to every hypothesis from which it can be derived. It isn't enough that every consequence provides some amount of support or other to all its parent hypotheses. (A *parent hypothesis* of a proposition is a hypothesis from which the proposition can be derived.) Suppose that T1 and T2 are empirically equivalent and that the datum e is a consequence of both T1 and T2. The assumption that e provides some support for both T1 and T2 still isn't enough to allow us to conclude that T1 and T2 are evidentially indistinguishable, for e may provide different amounts of support for each. Laudan and Leplin don't need to establish that some consequences are completely nonevidential. A difference in degree will do. This shields their example from the potential retort that the data on puberty do provide some evidence, albeit very weak evidence, for the hypothesis. Laudan and Leplin's point wouldn't be affected by the admission that this might be so. Having said that, I plan to restrict my discussion to Laudan and Leplin's special case of the more general issue—the case where some empirical consequences have a confirmational weight of zero. What I have to say about the special case can be modified to apply to the broader issue of arbitrarily different weights. But it isn't worth the trouble to do so.

Laudan and Leplin give us an example of an empirical consequence (the puberty of the scripture-reading boys of Lynchburg) that doesn't confirm the hypothesis from which the consequence is derived (that scripture reading causes puberty). Let's grant that the confirmational situation is as they describe it. Even so, there are two reasons why this example is insufficient to underwrite the conclusion that EE doesn't entail UD. First, the fact that some empirical consequences fail to confirm some of their parent hypotheses is compatible with the possibility that these consequences may fail to confirm *any* of their parent hypotheses. If this were so, then the argument from EE to UD would remain unaffected—for given two empirically equivalent theories T1 and T2, the fact that some consequence e lacks confirmatory potency would merely result in one and the same subtraction from the confirmational level of both T1 and T2, leaving them tied in believability. What Laudan and Leplin needed to show by example was that one and the same empirical fact can confirm some of its parent hypotheses and at the same time fail to confirm others of its parent hypotheses. I don't think this would be difficult to do. There are bound to be some hypotheses that would be strengthened by the observation that 1,000 scripture-reading teenagers in Lynchburg experienced puberty. Nevertheless, Laudan and Leplin can't be said to have achieved their aim without elaborating on their example in this way.

The second lacuna is going to be rather more difficult to fill. They also need to show that one and the same empirical fact may fail to confirm some *theoretical* hypotheses of which it's a consequence, while managing to confirm other theoretical hypotheses of which it's a consequence. Here's where the same question-begging move is made. Proponents of the underdetermination argument can accommodate themselves to Laudan and Leplin's example, even to the example that they should have given of an empirical fact that confirms some parent hypotheses but not others. They can accept this, but deny that there are any cases where this happens with theoretical hypotheses. They will meet Laudan and Leplin halfway on the question of theoretical hypotheses: they'll admit that it can happen that empirical facts fail to confirm parent theories. What they won't agree to is the suggestion that empirical facts ever *do* confirm their parent theories. But this essential piece of the argument is certainly not supplied by Laudan and Leplin's example, or by anything else that they have to say on the subject. Moreover, if they could supply the missing piece—if they could get their opponents to admit that some empirical facts confirm their parent theories—then they wouldn't need to go on with the argument. Realism would already have won. This is, of course, an exact repetition of what was said in section 6.2 about indirect confirmation. In fact, it's a pretty close repetition of what I've had to say about dozens of arguments on both sides of the realism debate: they're question-begging and redundant. I apologize for being tedious. But my material demands it.

Here's another, equivalent account of what's going on. Laudan and Leplin are trying to persuade us that empirical data confirm some parent hypotheses and fail to confirm other parent hypotheses. If they succeed, they show that the level of confirmation achieved by a hypothesis isn't always and completely determined by what its empirical consequences are. This fact had already been established by the validity of indirect confirmation, and we saw that it didn't do the job that Laudan and Leplin wanted done. So it's not surprising that another proof of the same proposition won't help.

Once again, this gambit lays the antirealist open to the charge of arbitrariness: why should one and the same property bear on the epistemic standing of one class of hypotheses but not another? I deal with that issue in due time (chapter 7). It should be appreciated, however, that the charge of arbitrariness is an entirely different argument from those that are mounted by Laudan and Leplin. If it's their arbitrariness that convicts the antirealists of irrationality, then we don't need to delve into the intricacies of indirect confirmation or nonconfirming consequences to make the point. The generic argument discussed in section 6.1 will do the job: antirealists evidently use some nonempirical criteria to make choices among empirical hypotheses, so what's the rationale for refusing to extend the same criteria to theoretical hypotheses? Laudan and Leplin's arguments don't make life any more difficult for antirealists than it already is.

6.4 A Note on Entity Realism

It would be unthinkable for a book that purports to canvass the debate about scientific realism to say nothing about entity realism (Cartwright 1983; Hacking 1983). The only thing I have to say about it, however, is that it doesn't change the dialectical balance between realists and antirealists. Entity realists accept that we have adequate grounds for believing that some theoretical entities exist, but they concede to antirealism that we're never justified in believing any theory about those entities. According to Hacking, we should (and do) believe in the existence of theoretical entities when we're able to manipulate them in order to investigate other, more speculative phenomena. We look for quarks by trying to induce fractional charges on a niobium ball by spraying it with electrons, and it's this engineering use of electrons that underwrites their reality: "If you can spray them, they are real" (Hacking 1983, 23).

There are, of course, two ways to understand the entity realist's slogan. The uninteresting way is to interpret sprayability (more generally, manipulability) in such a manner that it logically entails existence (how could you spray something that doesn't exist?). This interpretation invites the antirealist to question whether, in the hunt for quarks, electrons are *really* sprayed, or whether the operation described as "spraying a niobium ball with electrons" is itself to be given an instrumentalist reading. If the claim that manipulability entails existence is to be understood as an argument that's intended to persuade the opposition, then its premise must be interpreted in a manner that the opposition is willing to accept. So, the entity realist's claim is that if you can spray electrons, in a sense that *doesn't* presuppose their existence, then you have adequate but defeasible grounds for believing that they exist.

Now *this* claim strikes me as having all the shortcomings of the first and weakest of the three types of argument for the epistemic significance of nonempirical virtues enumerated in section 6.1: you cite a nonempirical property of theories (e.g., simplicity, explanatoriness) and claim that *it* has a bearing on our epistemic judgments. In the case of entity realism, you cite the manipulability of the theory's theoretical entities (in the sense that doesn't logically entail their existence) and claim that *it* licenses a realistic inference. One novelty of this form of realism is that the usual realist inference to "T is true" is weakened to "the theoretical entities posited by T exist". But I don't see that this attenuation changes the evidential relation between antecedent and consequent. The entity realist's conditional — "if you can spray them, they're real" — has all the weaknesses of the more traditional realist claims, such as "if it's the best explanation of the data, then it's true". In particular, neither conditional is motivated by an appeal to a broader principle that antirealists are compelled — or even inclined — to accept. In both cases, the desired conclusions are dished out as naked posits. The only consideration appealed to is the reader's intuition. No doubt appeals to intuition have their place in philosophical discourse. But a minimum condition on their suitability is that the intuitions deployed be shared by both sides in the debate. It seems clear, however, that antirealists who haven't been swayed by the epistemic claims of explanatoriness and simplicity are not going to be seduced into realism by allusions to manipulability. If antirealists believed in the epistemic equivalence of empirically equivalent theories because they had surveyed the available candidates for epistemic import and found each of them to be individually wanting, then calling their attention to a new property such as manipulability might have some impact. But antirealists' belief in premise 2 isn't based on this kind of enumerative induction over the available candidates. It's based on a predilection for the general principle that the epistemic import of a hypothesis is exclusively tied to its empirical content. If you're talking to people like that, there's no point in merely calling their attention to novel criteria of epistemic evaluation that they might not yet have considered. You need an argument that forces them to reevaluate their empiricist epistemology.

7

The Vulnerability Criterion of Belief

T he attentive reader will have noticed that none of the arguments against the second premise, that empirical equivalence entails underdetermination, has managed to survive the critical scrutiny of chapter 6. To my knowledge, there are no persuasive refutations of this thesis. Of course, the failure to show that a thesis is false
doesn't imply that it's true. This chapter begins with an examination of antirealist
attempts to convince us to adopt the second premise.

Actually, to speak of "attempts" in the plural is already to be overly sanguine about
the prospects for success. Combing the literature of antirealism turns up only two passages, both by van Fraassen, that might be construed as arguments for the second
premise, and the second of these passages merely supplies a part of the analysis that's
missing from the first. So there's really only one argument for the second premise on
the table. Moreover, this one and only argument has several strikes against it before
the analysis even begins. For one thing, both passages have the character of passing
remarks; to find a coherent argument in them requires a fair amount of speculative
reconstruction. For another, there are portions of the reconstructed argument that are
inconsistent with some of van Fraassen's other published opinions. Finally, there are
still other portions of van Fraassen's body of work which suggest that the second of the
two cited passages shouldn't even be construed as an attempt at rational persuasion.
To refer as I do to "van Fraassen's argument" is to take considerable liberties. The justification for doing so is that other philosophers have interpreted the passages in question as expositions of an argument. Their treatments establish at least that the argument I'm about to evaluate is one that's in philosophical play. At any rate, there's nothing
else to talk about under the heading of "arguments for the second premise."

7.1 The Deflationist Rejoinder

All of the arguments *against* the second premise discussed in chapter 6 seem to engender the same sort of dialectical interplay. In each case, realists cite a principle to
the effect that some property of hypotheses other than the identity of their empirical

consequences is relevant to their epistemic evaluation. This property is variously conceived as a nonempirical virtue like simplicity or explanatoriness, or as indirect confirmation, or as the diminished confirmational effect of some consequences on their parent hypotheses, or as the property that the theoretical entities posited by the hypothesis are manipulable. In some instances, antirealists may be able to reply that there's no non-question-begging reason that they should concede that the property in question is epistemically significant. This rejoinder isn't universally available, however. In particular, the case for indirect confirmation is utterly compelling. Moreover, even though antirealists may be able to deny the epistemic significance of any *specific* nonempirical virtue, they may have to accept the existential claim that there are *some* epistemically significant nonempirical virtues—or else they couldn't account for why they believe that future emeralds are more likely to be green rather than blue. But antirealists have another rejoinder available to them—one that disarms each and every realist argument against the second premise, regardless of whether or not the source of epistemic significance posited by the argument is accepted. In every case, antirealists can concede that the property in question may be epistemically significant, but deny that this source of epistemic significance *ever* helps to break epistemic ties among empirically equivalent theories. The fact that this line can be maintained without inconsistency shows that the realist arguments against the second premise don't have the force of a deductive proof of its falsehood from premises that the antirealist is required to accept. But now the question is whether antirealists are able to compel assent from the realists for their differential treatment of empirical and theoretical hypotheses.

Here's another way of formulating the question. Let R be a rule that ranks all the hypotheses that are consistent with the data in terms of their believability (I assume, as usual, that believability, or rationally warranted belief, is a probability function). R may take nonempirical properties such as simplicity or explanatoriness into account. It may also (or instead) allow for the effect of indirect confirmation. Whatever. Antirealists want to restrict the use of R to empirical hypotheses. Let R* be the rule whose dictates agree with those of R in all such cases, but that has nothing to say about the relative epistemic merits of theoretical hypotheses. We've seen that there's no logical impropriety in adopting R* but rejecting R. But why *should* we do it?

The argument begins with the claim that R* is to be preferred over R on the grounds of parsimony. R* is, after all, a weaker rule. There's actually no place in van Fraassen's writings, or in those of any other antirealist, where the point is put as baldly as that. But one can infer that this argument has been perceived in van Fraassen's work from the refutations that are offered by several philosophers. (This is rather like inferring the existence of dense nuclei from the scattering of alpha particles.) Leplin (1987), for example, objects that it's van Fraassen's antirealism itself that is "guilty of the inflationism it discerns in realism" (523). He continues:

> [Van Fraassen's antirealism] licenses standard inductive inference from evidence to belief so long as we deal in the realm of the observable, then suddenly blocks such inference in favour of a more modest epistemic stance once the boundary of the observable is reached. (523)

This passage makes it pretty clear that Leplin regards van Fraassen as having made the argument that we should stop at R* because moving to R would be inflationary.

A similar counterargument has been made by Arthur Fine (1986, 168). As far as I can tell, the target of these critical attacks must have been van Fraassen's often-quoted parable of the sheep and the lamb:

> There does remain the fact that . . . in accepting any theory as empirically adequate, I am sticking my neck out. There is no argument there for belief in the truth of the accepted theories, since it is not an epistemological principle that one might as well hang for a sheep as for a lamb. (1980, 72)

Now this passage doesn't contain an entirely explicit endorsement for stopping at R*. For one thing, van Fraassen is talking about theories rather than rules for evaluating theories. But the moral that van Fraassen wishes to draw from his parable is presumably going to be the same in both cases: by going from empirical adequacy to truth or from R* to R, we're sticking our necks out more than we would if we abided in empirical adequacy or R*. But is the moral the one we're looking for? In this passage, van Fraassen isn't claiming that there *is* an argument for stopping at R*. To do that, he'd have to say that it *is* an epistemological principle that one should *not* hang for a sheep when one can hang for a lamb. Strictly speaking, the parable of the sheep and the lamb merely claims that there is no argument for moving from R* to R (or from T* to T). This much is corroborated by the analysis in chapter 6. But it's clearly *suggested* that we ought to stick out our necks as little as possible—that is, that we ought to prefer R* on the grounds of epistemic safety. In any case, whether or not the argument can be attributed to van Fraassen, let's evaluate Leplin's critique.

Leplin's idea is that realism is more parsimonious because it licenses a single inductive rule, whereas van Fraassen's antirealism deploys two—one for observables and one for nonobservables. The appeal, as Leplin sees it, is to some notion of *simplicity*. Now I've already had occasion to rehearse the various ways in which one gets into trouble with epistemic appeals to simplicity. The relevant discussion occurs in section 5.8, where I examine McMichael's claim that T is simpler than T*. Most of what I have to say about McMichael's proposal applies to Leplin's as well. There's an additional problem with Leplin's argument, however: he's attributing an argument from simplicity to *van Fraassen*. In light of the fact that van Fraassen has made a career out of claiming that the nonempirical properties of hypotheses are merely pragmatic rather than epistemic virtues, this is an extremely implausible attribution.

But what about the parable of the sheep and the lamb? It seems to me that the greater parsimony claimed for R* doesn't have anything to do with its being syntactically or semantically or metaphysically simpler than R. If van Fraassen discerns "inflationism" in the realist's acceptance of R, it's because R* is *logically weaker* than R, in the straightforward sense that R logically implies R*, but R* doesn't imply R. Therefore, it must be the case that R* is more likely to avoid incorrect results than R. This is the sense in which accepting R is sticking one's neck out more than if one stopped at R*.

But to give this rationale for stopping at R* is to invite a recapitulation of the analysis in section 3.3. In that section, I examined van Fraassen's argument to the effect that antirealism is a weaker, hence, better account of scientific practice than realism. It was pointed out that the fact that one hypothesis is weaker than another isn't reason enough to repudiate either one. By the same token, the greater likelihood of R* doesn't provide a justification for the antirealist's selective skepticism. If this is what the deflationist rejoinder comes to, then it merely substitutes a new question for the one we started with: why we should accept R* but not R? The deflationist reply to this new question is that we should do so because R is epistemically riskier. But now the question becomes: why should we draw the line in the amount of risk that we're willing to take in such a manner that the risk entailed by R* is acceptable but the risk entailed by R isn't? After all, there are rules that are even weaker, hence epistemically safer, than R*. For instance, we could restrict our beliefs to what is actually going to be observed at some time in the history of the universe, rather than extend it to events that are observable in principle but will never be observed. This more severe restriction still involves sticking one's neck out, but the recommended degree of extension is less than that dictated by R*. Pointing to the fact that R* is weaker than R is not by itself an adequate defense of the second premise of the argument from underdetermination.

Moreover, despite the 1980 parable of the sheep and the lamb, I don't think that the van Fraassen of 1989 would endorse the deflationist reply. For the admission that R has a degree of believability, albeit a lesser one than R*, ultimately entails that we must be willing to ascribe discrete nonzero degrees of believability to theoretical hypotheses. But the 1989 van Fraassen holds the view that this admission destroys the case for constructive empiricism:

> Consider . . . the hypothesis that there are quarks. . . . The scientist has initially some degree of belief that this is true. As evidence comes in, that degree can be raised to any higher degree. That is a logical point: if some proposition X has positive probability, conditionalizing on other propositions can enhance the probability of X. (193)

In brief, it's van Fraassen's view that if we admit that claims about unobservables have low probabilities, then we can't resist the conclusion that there are circumstances under which these claims must be believed. But this is to concede that minimal epistemic realism is true. According to van Fraassen, antirealists should say that the probabilities of claims about unobservables are not low, but radically *vague* (1989, 194).

The foregoing point has been missed by several of van Fraassen's critics, who have supposed that his case for constructive empiricism is based on the extreme improbability of any particular theory's being true (Leeds 1994; Psillos 1996). It's easy to understand why they make this supposition. Van Fraassen's disquisition on the degrees of neck extension strongly suggests that he regards belief in theories to be riskier than belief in their empirical consequences, hence that belief in theories has a measurable degree of risk. To be sure, the relevant passage was written in 1980, whereas the repudiation of discrete theoretical probabilities occurs in 1989. But even in 1989, van Fraassen writes:

> I believe, and so do you, that there are many theories, perhaps never yet formulated but in accordance with all the evidence so far, which explain at least as well as the best we have now. Since these theories can disagree in so many ways about statements that go beyond our evidence to date, it is clear that most of them by far must be false. I know nothing about our best explanation, relevant to its truth-value,

> except that it belongs to this class. So I must treat it as a random member of this class, most of which is false. Hence it must seem very improbable to me that it is true. (146)

So it isn't surprising that some commentators have assumed that van Fraassen wants to ascribe very low probabilities to theories. But the fact remains that van Fraassen explicitly rejects the idea that theories may be ascribed discrete nonzero probabilities (1989, 193–194). As for the probabilistic argument quoted immediately above, it's quickly followed by a disclaimer:

> David Armstrong, replying to a version of this argument, writes "I take it that van Fraassen is having a bit of fun here." Yes, I had better own up immediately: I think I know what is wrong with the above argument. But my diagnosis is part and parcel of an approach to epistemology . . . in which rules like IBE [inference to the best explanation] have no status (*qua* rules) at all. As a critique of IBE, on its own ground, the above argument stands. (147)

The point is that *if* you're going to ascribe degrees of believability to theories on the basis of their explanatory virtues, *then*, van Fraassen claims, they have to be extremely low probabilities. The refutation of this argument would shield the *realist* from one of van Fraassen's attacks. But it wouldn't constitute a challenge to constructive empiricism, because van Fraassen doesn't believe that theories have discrete probabilities.

Incidentally, it isn't clear to me that van Fraassen has given a compelling reason for antirealists to refrain from ascribing low probabilities to theories. His claim is that if you're willing to attach a low probability to any theory, then you have to be willing to grant that there are scenarios in which this probability is elevated to "any higher degree" (1989, 193). It may be true that there are schemes for accommodating one's probabilities to incoming evidence that have this consequence. But this isn't to say that such schemes are obligatory for antirealists. In section 6.1, I discussed the option of setting a ceiling of $1/n$ to the probabilities that can be assigned to n mutually exclusive and empirically equivalent theories. Since the theories are empirically equivalent, we can be sure that no incoming evidence will ever *force* us to favor any one of these theories over the others. A coherent antirealist position can therefore be put together out of the foregoing probabilistic principle, along with the belief that every theory has empirically equivalent rivals. Of course, to say that this makes for a coherent position is not yet to show that this position is superior to its rivals.

In any event, the issue of whether theories have probabilities doesn't affect the main conclusion of the present section. It's explicitly van Fraassen's view that antirealists can't afford to ascribe degrees of risk to theoretical beliefs. This principle could be turned against the deflationist rejoinder that some people also attribute to van Fraassen: to claim that R is riskier than R* is to concede that R has a measurable degree of risk, which entails that belief in theories has a measurable degree of risk; therefore, if you don't believe that theories have degrees of risk, then you can't say that R is riskier than R*. In the preceding paragraph, I expressed some doubt about the incompatibility between antirealism and the ascription of degrees of risk to theories. If I'm right, then one potential source of trouble for the deflationist reply is averted. But the deflationist reply is severely defective regardless of whether antirealists are allowed to ascribe degrees of risk to theories and methodological rules. The more basic problem is that one hypothesis being weaker than another is insufficient ground for repudiating either one. If the existence of weaker states of opinion were a sufficient reason for rejecting a hypothesis, intellectual discourse would be reduced to an unrelieved exchange of tautologies. To assert anything more would be to stick one's neck out more than one might. So, now the question is: why should we stick our necks out to the extent of R* and no further? Unless this question receives an answer, the deflationist rejoinder merely trades one form of arbitrariness for another.

7.2 The Vulnerability Criterion of Belief

Why should we stop at R*? There's at least a suggestion of an answer in a 1985 article by van Fraassen. The following passage was quoted in chapter 3, but we need it here again:

> If I believe the theory to be true and not just empirically adequate, my risk of being shown wrong is exactly the risk that the weaker, entailed belief will conflict with actual experience. Meanwhile, by avowing the stronger belief, I place myself in the position of being able to answer more questions, of having a richer, fuller picture of the world, a wealth of opinion so to say, that I can dole out to those who wonder. But, since the extra opinion is not additionally vulnerable, the risk is—in human terms—illusory, and *therefore so is the wealth*. It is but empty strutting and posturing, this display of courage not under fire and avowal of additional resources that cannot feel the pinch of misfortune any earlier. What can I do except express disdain for this appearance of greater courage in embracing additional beliefs which will *ex hypothesi* never brave a more severe test? (255)

There's an exegetical problem attaching to this passage. Its date puts it squarely between *The Scientific Image* (1980) and *Laws and Symmetry* (1989). Now, there are some important philosophical differences between van Fraassen's defense of antirealism in 1980 and his defense in 1989. I discuss these differences at length in chapter 12. For present purposes, it suffices to characterize them as follows. In 1980, van Fraassen believes that he's presenting a case for constructive empiricism that rationally compels assent from the realists. In 1989, he concedes that there's no rational way to compel assent from the realists, but maintains that the espousal of antirealism is nevertheless not irrational. Coming as it does midway between the hard line of 1980 and the softer line of 1989, the passage from 1985 can be interpreted either way. The question here is whether those for whom the 1985 van Fraassen expresses disdain are being accused, à la 1980, of failing to adopt a compelling epistemological principle or, à la 1989, merely of bad taste. Viewed in the retrospective light of his 1989 discussion, the latter is a very reasonable interpretative hypothesis. But it's also possible to interpret the intermediate van Fraassen as attempting to provide the missing warrant for the 1980 arguments. Certainly some commentators have treated the passage quoted above as an argument rather than an attitude (e.g., Clendinnen 1989). Whichever is the correct interpretation, it's worth investigating whether there are any objective grounds for van Fraassen's disdain.

What does this disdain look like if it's elevated to the status of an epistemological principle? The provisional answer given to this question in chapter 3 was good enough for the purpose at that time. But it wasn't quite right. Let's say that two hypotheses T1 and T2 are *equivalently vulnerable* if there's no possible observation that disconfirms one of the hypotheses but not the other. Van Fraassen's principle seems to be that if T1 and T2 are equivalently vulnerable, and if T1 is logically stronger than T2, then we should not believe T1. Let's call this the *vulnerability criterion of belief* (VCB). At first glance, equivalent vulnerability seems to come to the same thing as empirical equivalence. If that were so, then the provisional interpretation of van Fraassen's disdain given in chapter 3 would come to the same thing as VCB. But a more careful examination reveals some hidden complexities.

Note, to begin with, that VCB is strongly reminiscent of the logical positivists' Verifiability Criterion of Meaning (VCM). Like VCM, it's presumably intended to apply only to contingent statements, and not to a priori truths. Like VCM, it's susceptible to Quinean attacks based on holism that threaten to reduce it to vacuity. However, VCB is logically *weaker* than VCM, for meaningless propositions can't be believed, whereas the unbelievable may still be meaningful. Evidently, VCB plays the same gatekeeping role in a diminished neo-empiricist theory of knowledge that VCM once played in the logical positivists' stronger brand of empiricism.

Upon superficial examination, VCB seems to provide the missing warrant for refusing to break epistemic ties among empirically equivalent theories — equivalently, for refusing to go from R* to R. For suppose that T1 and T2 are empirically equivalent theories. Let T* be the hypothesis that T1 (or, just as well, T2) is empirically adequate. Now suppose it's true that empirical equivalence entails equivalent vulnerability. Then, since T* is empirically equivalent to T1 and T2, it's also equivalently vulnerable to T1 and T2. But T1 and T2 are both logically stronger than T*. Therefore, by VCB, we may not believe either one. And therefore this crucial premise for the underdetermination argument is secured. The present section started with a question about the deflationist reply: why should we accept the degree of neck extension that attaches to R* but not the greater degree that attaches to R? VCB answers: because the "extra opinion" that comes with R is not "additionally vulnerable".

But VCB provides a great deal more than the missing warrant for the second premise of the underdetermination argument. In fact, it's arguably the core thesis on which the whole case for epistemic antirealism depends. (That's why the title of this chapter omits any limiting reference to underdetermination.) In section 3.3, I showed that an appeal to the as yet unnamed VCB provided the missing warrant for van Fraassen's *scientific practice* argument: if both realism and antirealism can account for any scientific practice, and if it's true that the antirealist account is logically weaker, then VCB seems to purchase the conclusion that we should be antirealists. Section 3.3 also made the point that the line of reasoning obtained by joining the scientific practice argument with VCB is unnecessarily convoluted. With VCB in hand, there seems to be a quick and easy argumentative path to the desired antirealist conclusion that circumvents the issues relating to scientific practice. For any theory T, let T* be the hypothesis that T is empirically adequate. Now, T* is obviously weaker than T. Moreover, belief in T doesn't render us more vulnerable to experience than belief in T*. Therefore, by VCB, we should at most believe in T* — which is to say that we should be antirealists. This is the argument that I called GA in section 3.3. Note, once again, that GA makes no reference to scientific practice. Now note additionally that it also makes no reference to the indefinitely many empirically equivalent rivals needed for the underdetermination argument. On the other hand, both the argument from underdetermination and the argument from scientific practice must presuppose VCB, or something equally potent, to arrive at their conclusion. Thus, the 1985 defense of constructive empiricism actually renders the 1980 arguments superfluous. This shields constructive empiricists from a number of potential difficulties. If their case depended on the underdetermination argument, they would have to worry about attacks, such as Laudan and Leplin's (1991), on the premise that every theory really does have indefinitely many empirically equivalent rivals; and if it depended on the argument from scientific practice, they'd have to worry about how to make the distinction between the essential scientific practices that antirealism has to account for, and the inessential practices, such as wearing white lab coats, that can safely be ignored. With the quick and easy argument GA, however, these problems and pitfalls become irrelevant. GA seems to get closer to the heart of van Fraassen's reasons for antirealism than do any of the 1980 arguments. Both van Fraassen's defenders and his critics would have been spared a great deal of trouble if GA, and the vulnerability criterion it relies on, had been enunciated in 1980.

7.3 Conjunctive Empiricism Revisited

Unfortunately for constructive empiricism, the quick and easy argument doesn't work, even if the validity of VCB is granted. If what it means to be a neo-empiricist is to subscribe to VCB, then neo-empiricists should not be constructive empiricists—at least, not if "constructive empiricism" is interpreted narrowly as the precise philosophy of science expounded in *The Scientific Image*. For there are ways in which belief in theories makes us vulnerable over and above belief in their empirical adequacy. Here is one way. Suppose we believe that theories T_1 and T_2 are true, and that E is an empirical consequence of T_1 & T_2 but not of either T_1 or T_2 in isolation. Now our belief in T_1 and in T_2 is vulnerable to disconfirmation by the possible empirical discovery that E is false. In contrast, if we believe only that T_1 and T_2 are both empirically adequate, then E may very well not be a consequence of our beliefs. Therefore, belief in T_1 and in T_2 is vulnerable in ways that belief in their empirical adequacy isn't. This argument should not be mistaken for Putnam's conjunction argument for realism. The latter is a species of scientific practice argument which claims that the prevalent practice of believing in the empirical consequences of conjoint theories *presupposes* realism. Van Fraassen's 1980 response to this argument was to deny that theories ever are conjoined. In chapter 3, I expressed the opinion that Trout (1992) has shown van Fraassen to be wrong on this point. Be that as it may, the force of the present counterargument against GA doesn't depend on whether van Fraassen is right or wrong about conjunction. Regardless of whether scientists conjoin theories, or whether they should conjoin theories, the fact remains that full belief in T_1 and T_2 makes us more vulnerable than belief in their empirical adequacy.

   The foregoing argument shows that constructive empiricism, as van Fraassen conceives it, isn't the philosophy of science dictated by VCB. But it doesn't yet dis-

pose of the general thesis of epistemic antirealism. For the vulnerability due to theoretical conjunction is entirely compatible with an old philosophical friend, *conjunctive empiricism*. Let's review the difference between T* and T#. For any theory T,
T* is the thesis that the empirical consequences of T are true, whereas T# is the stronger thesis that asserts the truth of all of the empirical consequences of T in conjunction with any of the currently accepted auxiliaries. (See section 3.1 for more detail.)
Constructive empiricism in its narrowly van Fraassian sense is the view that we're
never warranted in believing more than T*. Conjunctive empiricism is the thesis
that we're never warranted in believing more than T#. Conjunctive empiricists
allow themselves a greater epistemic latitude than their constructive empiricist cousins. But they're still antirealists. Their willingness to believe T# doesn't underwrite
the belief that the theoretical entities postulated by T exist. The punchline is that T#
is already vulnerable to disconfirmation by empirical facts that contradict the consequences of T in conjunction with other accepted theories. This particular vulnerability is therefore not a counterexample to the philosophical claim that VCB
entails antirealism.

It's instructive to make the same point in a slightly different way. Let $T_1, T_2, \ldots,$
T_n be the list of all the theories that we currently find acceptable. $T_1 \& T_2 \& \ldots \& T_n$
is therefore our current *total science*. Realists believe that all the T_i are true. Constructive
empiricists believe that $T_1^*, T_2^*, \ldots, T_n^*$ are true. Conjunctive empiricists believe
that $T_1\#, T_2\#, \ldots, T_n\#$ are true. But for any i, $T_i\#$ is the same as $(T_1 \& T_2 \& \ldots \&$
$T_n)^*$. That is to say, the conjunctive-empirical consequences of any accepted theory
are identical to the empirical consequences *tout court* of our current total science.
(The difference between empirical consequences and conjunctive-empirical consequences collapses in the case of total sciences.) Thus, conjunctive empiricism is
just the constructive-empiricist attitude applied to total sciences. Now, we've already
encountered reasons for reformulating constructive empiricism as a thesis about total
sciences. In section 5.2, I discussed Laudan and Leplin's argument against the *first*
premise of the underdetermination argument—the assumption that every theory has
empirically equivalent rivals. I showed that this argument could be circumvented in
the case where the theory is a total science. Fortunately for antirealism, this is enough
of an admission to get the underdetermination argument off the ground. Though I
didn't say so in section 5.2, the reformulation of constructive empiricism as a doctrine about total sciences makes it identical to conjunctive empiricism. So, conjunctive empiricism solves at least two problems for antirealists.

Here is a third reason for reformulating constructive empiricism as a thesis about
total sciences. Individual, as opposed to total, theories are generally composed of a
number of logically independent postulates. Evidently, constructive empiricism, in
its unregenerated van Fraassian form, has no compunction about believing in the
empirical consequences of the conjunction of these postulates. But, given a number
of independent theoretical claims, there is no principled way to decide whether they
comprise a collection of several theories, or of several postulates of a single theory.
Thus, constructive empiricism is committed to the belief in the empirical consequences of conjoined theoretical claims right from the start.

What I'm trying to say, in brief, is that there are a lot of independent reasons
why antirealists should move from strict constructive empiricism to conjunctive

empiricism. One of the payoffs of doing so is that they would no longer be suscep-tible to the VCB-based objection that we began this section with. Disconfirmation by conjoint-empirical consequences doesn't make realist believers in T more vul-nerable than antirealist believers in T#.

7.4 T*, T#, T^, and T

But there's a way in which belief in T makes us more vulnerable to experience than either belief in T* or belief in T#. This vulnerability has been discussed by Boyd (1984) and somewhat later but apparently independently by Clendinnen (1989). If we believe T to be true, then we also believe that *the empirical consequences that will be derivable from the conjunction of T with the as yet unformulated theories that will be accepted in the future* are also true. This belief is vulnerable to future experi-ence in a way that mere belief in the empirical adequacy of our present theories — even of our present total science — is not. Indeed, the vulnerable content of this new belief can't be represented as a hypothesis about the empirical adequacy of any theo-retical proposition, for the simple reason that we don't yet know what theories will be formulated and accepted in the future. The conjunction objection of section 7.3 forces constructive empiricists into saying that our warranted beliefs can only be subsumed by the claim that our total science is empirically adequate. But this doc-trine is still very close to van Fraassen's original version of antirealism. Van Fraassen might very well regard it as nothing more than a friendly amendment. However, the vulnerable belief in the empirical adequacy of the conjunction of our current theo-ries with theories yet to be formulated constitutes a more radical departure. This form of vulnerability establishes that neither constructive nor conjunctive empiricism is dictated by VCB.

Boyd and Clendinnen think that this argument against constructive/conjunc-tive empiricism can be extended into an argument for realism. In this, however, I think they go too far. The conjunction argument of section 7.3 showed that construc-tive empiricism needed to be broadened to allow for more than belief in T*. But it was shown that the vulnerability due to conjunction could be accommodated by expanding one's domain of potential beliefs from T* to T#. There was no need to go all the way to T. Boyd's and Clendinnen's new form of vulnerability is experienced neither by believers in T* nor by believers in T#. Thus, VCB allows us to broaden our epistemic range beyond T* and T#. But we still don't have to go all the way to T. The new form of vulnerability can be accommodated by a willingness to believe in T^, which asserts nothing more than that the empirical consequences of T in con-junction with future auxiliaries will be true. T^ can't be derived from T#, much less from T*. It's a distinct hypothesis of its own. Let's call the view that allows for belief in T^ by the name of *prospective empiricism*. Prospective empiricists allow them-selves to have beliefs that neither constructive empiricists nor conjunctive empiri-cists would entertain. But they're still not realists.

As I've defined them, T# and T^ are independent claims: the former says that the empirical consequences of T in conjunction with the *present* auxiliaries are true; T^ says that the empirical consequences of T in conjunction with *future* auxiliaries will turn out to be true. But it would be very eccentric to believe T$^\wedge$ but not T#. I therefore redefine T$^\wedge$ as the view that the empirical consequences of T in conjunction with both present and future auxiliaries are true. In this sense, T$^\wedge$ entails T#. T*, T#, and T$^\wedge$ comprise a sequence of transformations of T in which each item is logically entailed by its successor.

But is T$^\wedge$ coherent? After all, the auxiliaries of the future may very well lead to empirical predictions that contradict the predictions that can be made on the basis of the present auxiliaries. How, then, can one suppose that the consequences of T in conjunction with either one are true? Well, you could produce the same dilemma for the full-fledged realist belief that T is true. If we believe that T is true, then we believe in the empirical consequences of T in conjunction with any other theory that we currently believe, and we also believe in the empirical consequences of T with theories that we'll *come* to believe in the future. There is no dilemma here because part of believing that our current theories are true is believing that the theories that will be discovered in the future will *not* be inconsistent with them. The same can be said about belief in T$^\wedge$. This point obviously needs more clarification. I'm going to refrain from developing it further, however, because the coherence of T$^\wedge$ isn't essential to my argumentative agenda. My claim is that belief in T$^\wedge$ renders one just as vulnerable as belief in T, so VCB still dictates an antirealist stance. If it should turn out that T$^\wedge$ is incoherent, it would simply mean that there is no additional form of vulnerability to consider—and then I wouldn't *need* to make any claims about T$^\wedge$ to conclude that VCB dictates an antirealist policy.

Both Boyd and Clendinnen explicitly consider the merits of prospective empiricism (although not by that name) as an alternative to realism. Boyd's argument against stopping at T$^\wedge$ is straightforwardly that it's only the truth of T that provides us with an *explanation* for why a hypothesis such as T$^\wedge$ might be true. But this is the question-begging miracle argument all over again. In the original miracle argument, the claim is that the truth of T explains the truth of T*; here the argument is that the truth of T explains the truth of T$^\wedge$. Needless to say, those who remain unconvinced by the first miracle argument are not going to buy the second. On the other hand, those who buy into the second are also going to accept the first and so will have no need of the second. This is another argument that, right or wrong, doesn't do anybody any good.

At first glance, Clendinnen's argument for going all the way from T$^\wedge$ to T seems to add a few new wrinkles. Clendinnen notes that T$^\wedge$ is *parasitic* on T, in the sense that T$^\wedge$ can only be described in terms of T. He then argues that the parasitism of T$^\wedge$ compels us to believe in T if we believe in T$^\wedge$.[1] Actually, Clendinnen's discussion doesn't clearly differentiate T$^\wedge$ from the weaker putative parasite, T#. Running the argument on T$^\wedge$ rather than T# strengthens it considerably, however—for two reasons. First, since T$^\wedge$ entails T#, any argument to the effect that belief in T# commits us to belief in T is only going to have its premise strengthened by the substitution of T$^\wedge$ for T#. The distance between T$^\wedge$ and T that needs to be bridged by the argument is smaller than the distance between T# and T. Second, Clendinnen has a better chance of making the parasitism charge stick to T$^\wedge$ than to T#. In chapter 5, I argued that it was by no means obvious that T* is essentially parasitic on T. The same arguments would apply as well to T#, which is nothing more than the *-transform of the total science of which T is a part. But the same arguments definitely don't apply to $T^\wedge$. There's at least a possibility that we might be able to give a finite axiomatization of the empirical consequences of T that avoids making any reference to T itself. This would show that T^* isn't parasitic on T after all. The same can be said of the conjoint-empirical consequences that comprise the content of T#. But there's no chance of pulling the same trick with $T^\wedge$. $T^\wedge$ says (in part) that the empirical consequences of the conjunction of T and the auxiliaries of the future are going to turn out to be true. There's no possibility of our being able to give an independent characterization of empirical consequences that comprise $T^\wedge$ *because we don't yet know what these empirical consequences are going to be*. There's no way to describe $T^\wedge$ except as the empirical consequences of the conjunction of T and the auxiliaries of the future. So, the claim that $T^\wedge$ is parasitic on T is on a much firmer footing than the corresponding claims for T^* and T#.

Clendinnen's argument for moving from $T^\wedge$ to T begins on a promising note:

> It is sometimes suggested that because two hypotheses both predict the same facts, they both explain these facts. The notion of goodness of explanation might be invoked to distinguish between them; but, as we have seen, it can then be argued that these criteria are not empirically significant. (1989, 84)

The last remark suggests that Clendinnen isn't going to present us with yet another question-begging appeal to the epistemic significance of a nonempirical theoretical virtue. But this is just what he does. He claims that "it is reasonable to accept" a principle which asserts that "if a hypothesis K can only be used to predict by first employing hypothesis H and then advancing just these same predictions, then H is thereby the better explanation of the facts predicted" (84–85). This principle does indeed conform to my intuitions about explanatory goodness. But it still doesn't purchase the conclusion that we should *believe* H without begging the question of whether the nonempirical virtues of theories like explanatory goodness should count as reasons for belief. Clendinnen continues:

> The predictive content of $[T^\wedge$ & not-T] is parasitic on [T], and this bears on the rationale we could have for accepting or believing the former. If we doubt the truth of [T] we have no basis to expect the predictions made by it. . . . The rationale for our predictions is the confidence we have in our speculation about hidden structures. $[T^\wedge]$ might be marginally more probable than [T], but in doubting the latter we abandon all grounds for confidence in the former. . . . $[T^\wedge]$ does not open up a route for predicting which sidesteps a decision about accepting [T]. . . . We remain obliged to make a decision about accepting [T] for the purpose of prediction. (86–87)[2]

The claim that "if we doubt the truth of [T] we have no basis to expect the predictions made by it" is, of course, the miracle argument once again. It's true that "$[T^\wedge]$ does not open up a route for predicting which sidesteps" *reference* to T, but it certainly opens up a route for predicting that sidesteps "a decision about accepting [T]" (I presume here that accepting T means believing it, or there would be no disagreement with van Fraassen). There's nothing new in this discussion that would impel a prospective empiricist to move from $T^\wedge$ to T. According to VCB, asserting the surplus of T over $T^\wedge$ is empty strutting.

It may be that there's another, still undiscovered way in which belief in T renders us more vulnerable than any belief that falls short of T. The burden of proof is on the realists to come up with a viable candidate. Until they do, the presumption will have to be that VCB dictates an antirealist stance, albeit one that is in some respects more liberal than strict constructive empiricism.

7.5 The Overly Liberal Element in Constructive Empiricism

There are also elements of constructive empiricism that are *too* liberal from the viewpoint of VCB. Let T be an empirical generalization such as "All emeralds are green". In the spirit of van Fraassen, let's decompose T into the conjunction To & Tn, where To is the theory which asserts that all emeralds that will ever be observed are green, and Tn is the theory which asserts that all the emeralds that will *never* be observed are green. Now consider To as a rival to T. On van Fraassen's notion of empirical equivalence, To is *not* empirically equivalent to T, since T has various implications about events that are observable but that will never be observed. Nevertheless, To is equivalently vulnerable to T, because we can be sure on a priori grounds that we will never encounter an experience that is relevant to the truth of T but not to To. The belief that even the emeralds that will never be observed are green is invulnerable. Therefore, by VCB, our belief should only extend to To. More generally, let T be a theory that may posit unobservable entities and processes. Constructive empiricism maintains that we should restrict our belief to T^*. The reason for this restriction is supposed to be that belief in the content of T over and above T^* is invulnerable. Now, all the claims of T^* are about observable events. Some of these events, however, will actually be observed, while others will not, simply because there will be no observer in the right place at the right time. T^* can be partitioned into $(T^*)o$ and $(T^*)n$, depending on which of these two types of events it deals with. Clearly, extending our belief from $(T^*)o$ to $(T^*)n$ does not make it more vulnerable. It's another empty strut.

Combining the present conclusion with that of section 7.4, we obtain the result that VCB leads to a position that is substantially different from van Fraassen's. According to constructive empiricism, we should restrict our belief to T^*. But VCB tells us that we should restrict our belief to $(T^\wedge)o$, which includes some claims that T^* excludes, and excludes some claims that T^* includes. The view that we should restrict our beliefs to $(T^\wedge)o$ can be called *restricted prospective empiricism*.

7.6 The Status of the Vulnerability Criterion of Belief

We've seen that there are some incompatibilities between the dictates of VCB and the specifically van Fraassian version of antirealism. The more important question is whether the appeal to VCB gives the general position of antirealism any additional leverage over realism. As far as I can ascertain, there's no vulnerability to which realist believers in T are heir that isn't shared by antirealist believers in $T^\wedge$. In fact, even the more skeptical believers in $(T^\wedge)o$ are vulnerable to anything that realists are. Thus,

VCB does seem to underwrite some version or other of antirealism. If there's a way to compel realists to adopt it, then antirealism wins. But why should realists adopt it?

The only reason that van Fraassen gives for the acceptance of VCB is that its violation doesn't *buy* you anything—"the risk is . . . illusory, and therefore so is the wealth" (1980, 255). But, of course, believing in empirically adequate theories doesn't *cost* you anything, either, so the economic argument fails to settle the issue. On the basis of introspection, I think that the lure of VCB is due to an obscure sense that nonbelief is a *default* position for rational beings—that we abide in the state of nonbelief unless we encounter persuasive reasons that impel us to move. Perhaps we think that some sort of law of least effort is involved. To articulate this notion is already to see that it's without foundation. At the moment we become reflective about our mental life, we already find ourselves immersed in a tangled net of beliefs. That is our de facto default position, and if we follow a law of least effort, we'll abide in our prereflective beliefs until circumstances force us to give them up. It's true that no experience can compel us to adopt an invulnerable belief about the world. But it's also true that no experience can compel us to abandon one that we already have.

Gilbert Harman (1986) gives additional reasons for adhering to the same conservative principle of belief revision. Harman claims that it isn't irrational to hold on to our current beliefs even if we no longer have any idea what the original justification for adopting them was. The reason, in brief, is that any other procedure would be computationally intractable: we simply can't keep track of the justificatory basis of every item in our knowledge base. According to Harman, we should adopt the "principle of positive undermining", which stipulates that we must stop believing P only when we positively come to believe that our reasons for believing P aren't any good. The conclusion of the preceding paragraph follows forthwith: the invulnerability of our beliefs in empirically adequate theories merely insulates them from the danger of having to give them up on the basis of experience.

Most important, the philosophical inclination to realism is itself grounded on the intuition that VCB is false. What turns people into scientific realists above all else is their belief that there are nonempirical properties of theories, such as their simplicity or explanatoriness, that have a bearing on their epistemic status. This belief is tantamount to the negation of VCB. To be sure, we've seen that realists have been unable to find any rational means of compelling antirealists to accept their intuition. But to base the argument *for* antirealism on VCB is simply to stipulate that the realists are wrong.

In sum, the case for the vulnerability criterion of belief has yet to be made. Since it's involved in both the arguments from underdetermination and the scientific practice argument for antirealism, it follows that these arguments, too, are inconclusive at best. On the other hand, none of the arguments for realism—the miracle argument, the conjunction argument, Friedman's argument—has met with success, either. The impasse continues.

8

The Belief-Acceptance Distinction

In chapters 8–11, I take up a final category of realist argumentation. These arguments try to show that scientific antirealism is an incoherent position. The main charge of this type is the claim that there's no coherent way to distinguish between the theoretical and the empirical consequences of a theory. If this is so, then antirealism is obviously a nonstarter. The examination of this claim takes up all of chapters 9–11. In this chapter, I deal with an argument of Horwich's (1991) to the effect that there can be no coherent distinction between believing in theories and the weaker attitude of "acceptance" recommended by constructive empiricists.

Van Fraassen (1980) tells us that scientists' *acceptance* of theories should not be construed as entailing that they *believe* those theories to be true. To be sure, the phenomenon of scientific acceptance involves the willingness to use the theory in deriving predictions about the future. But this use requires no more than a belief in the theory's empirical adequacy. Nor do the other phenomena related to acceptance — for example, a commitment to relying on the theory to fill requests for explanations — require full belief. Horwich claims that this position is incoherent because there's no difference between van Fraassian acceptance and belief. He considers four attempts to draw a belief-acceptance distinction and argues that they all fail. My point is going to be that, in the course of one of his arguments, Horwich concedes enough for constructive empiricists to formulate their position in a coherent manner. Given Horwich's view of the matter, the worst charge that can be laid against van Fraassen is that he violates common linguistic usage. His substantive epistemic claims, however, can be reformulated in a way that defuses Horwich's criticism.

The argument that Horwich criticizes is itself due to van Fraassen (1985). The latter notes that theoretical unification results in a theory that must be considered *less believable* than the conjunction of the theories that are unified, because it claims more about the world. However, the unified theory is *more acceptable* than the conjunction because it's simpler and therefore more useful. But if belief and acceptance respond differently to theoretical unification, they can't be the same thing. Horwich's counterargument is that van Fraassen implicitly assumes that *epistemic* values (e.g., consistency and conformity to the data) apply only to belief, whereas *pragmatic* values apply only to acceptance. But, in fact, we sometimes use pragmatic considerations in evaluating the appropriateness of a belief (as in Pascal's wager), and epistemic considerations are also relevant to acceptance. Thus, no difference between acceptance and belief has as yet been formulated. Horwich writes:

> Now, one could deny this. One could insist that considerations of abductive epistemic rationality do not apply to the mental state of acceptance, but this response would beg the question. It would simply presuppose that belief and acceptance are distinct states, and provide neither an argument for their distinctness nor a response to our grounds for skepticism on this point. In short, the difference between epistemic and pragmatic norms of evaluation does not entail a corresponding difference between states of mind to be evaluated. (1991, 7–8)

Actually, it's difficult to imagine that anyone, van Fraassen included, would deny that epistemic norms are relevant to rational acceptance. What van Fraassen does seem to deny is that pragmatic norms are relevant to rational belief. The rest of Horwich's remarks are unaffected by this substitution, however.

What are we to say of the claim that acceptance and belief respond differently to pragmatic considerations? Horwich charges that this claim merely begs the question. It seems to me that it's pointless to argue about whether belief is affected by pragmatic norms. Clearly, if we're capable of both epistemic and pragmatic evaluations, then we're capable of undertaking these evaluations either separately or jointly, as we deem advisable. There is a real question about whether the folk-psychological concept that goes by the name of "belief" refers to a mental state that is properly evaluated on epistemic grounds alone, or whether "belief" is evaluated on both epistemic and pragmatic grounds. But the answer to this question doesn't seriously affect the realism-antirealism issue. If "belief" is to be evaluated on epistemic grounds alone, then van Fraassen's distinction between belief and acceptance stands, and his position is coherent. If, on the other hand, "belief" may be evaluated on both epistemic and pragmatic grounds, then van Fraassen needs to reformulate his stance. Certainly antirealists like van Fraassen are not interested in denying that theories may have pragmatic value—quite the contrary! Thus, if Horwich is right about the proper use of the word "belief", it follows that constructive empiricists are guilty either of an embarrassingly elementary logical error or of violating common usage. An argument purporting to show that constructive empiricism is incoherent would have to show that the doctrine can't be saved by any simple hypothesis concerning what constructive empiricists really have in mind that they mistakenly refer to as "belief". But there is such a hypothesis. Constructive empiricists need only to introduce the notion of an *epistemic belief*, which is just like a belief, except that pragmatic considerations don't play a role in its evaluation. There can be no doubt that we're capable of epistemic beliefs, since the cognitive resources needed for epistemic-believing are a proper subset of those needed for forming and evaluating garden-variety beliefs. Nor can there be any doubt that epistemic belief is different from acceptance, for the latter is affected by pragmatic considerations whereas the former is not. But the distinction between epistemic belief and acceptance is enough to formulate the constructive empiricist viewpoint in a coherent manner: constructive empiricists maintain that there may be good reasons for accepting scientific theories, but never for epistemic-believing more than their empirical consequences.

In the passage quoted earlier, Horwich maintains that the difference between epistemic and pragmatic evaluation doesn't warrant the claim that there's a "corresponding difference between states of mind to be evaluated". He doesn't make explicit how states of mind are to be individuated in this context. Certainly the state of mind corresponding to having performed an epistemic evaluation differs from the mental state corresponding to a pragmatic evaluation in *content*, if nothing else. Perhaps Horwich intends to individuate mental states by the type of *propositional attitude* involved. This is the sense in which believing-that-P and desiring-that-P are different states. Now, it's true that the difference between epistemic and pragmatic evaluation doesn't entail the existence of any differences among propositional attitudes. It's also true that van Fraassen's discussion of belief and acceptance presupposes that these terms refer to different propositional attitudes. Very roughly, van Fraassian belief is a judgment, while van Fraassian acceptance is a commitment to a course of action. But, once again, these issues are tangential to the question of whether constructive empiricism is coherent. As already noted, we're free to *stipulate* that belief, or at least epistemic belief, is responsive only to epistemic considerations, whereas acceptance is also affected by pragmatic concerns. There is no question-begging here: the conceptual apparatus needed to make the distinction is supplied by Horwich himself.

On the other hand, we can't simply stipulate that belief and acceptance are different propositional attitudes. If they *are* different propositional attitudes—for example, if belief is a judgment and acceptance is a commitment to action—then the case for the coherence of constructive empiricism is made. But Horwich is right in claiming (if this is indeed his claim) that the difference between epistemic and pragmatic evaluation doesn't entail that there are two correspondingly different propositional attitudes. So let's suppose that they're the *same* type of propositional attitude. Perhaps they're both species of judgment: to believe a theory T is to judge that it has very high epistemic value, and to accept T is to judge that it has high pragmatic value, or something of the sort. This difference in the *content* of the two judgments is already enough to formulate a coherent constructive empiricism. The constructive empiricist is simply one who maintains that theories may properly be judged to have high pragmatic value, but never high epistemic value. To see that this formulation does the job, consider the following scenario. Suppose it were agreed all around that acceptance is indistinguishable from belief. According to Horwich's analysis, this would mean that constructive empiricism is incoherent. But suppose that it were also agreed that the only grounds for believing (=accepting) theoretical statements are pragmatic, whereas there are epistemic as well as pragmatic grounds for believing (a.k.a. accepting) a theory's empirical consequences. Since belief is identical to acceptance, it follows that theories *may* rationally be believed (because of their pragmatic value), which is the negation of the constructive empiricist claim. Yet this state of affairs would surely be regarded by constructive empiricists as a vindication of their point of view.

But what if we regard both belief and acceptance as species of commitments to action? This is, I think, the strongest tack for Horwich to take. It might be suggested that talk of "judgment" as a process divorced from action is archaic and illegitimate, and that any proposed distinction between belief and acceptance must be based on some actual or potential differences in what people are prepared to *do* in the corresponding states. If this line is correct, then the constructive empiricist is committed to finding a behavioral difference between what people might do when they actually believe a theory and what they might do when they merely accept it. Horwich's claim seems to be that such differences don't exist. My reply is that, even if belief and acceptance are both regarded as commitments to action, their differentiation doesn't depend on their being commitments to *different* actions. Believing T and accepting T may be different mental states even if they're associated with exactly the same behavioral dispositions. This is true of any two mental states. For example, we might be disposed to act in a certain manner because we want to conform to the dictates of society, whatever the shortcomings of those dictates may be; or, we could be disposed to act in the very same manner because we think that social norms, embodying as they do the wisdom of collective experience, are the best guide to action. The identity of behavioral dispositions doesn't entail the identity of mental states. Of course, it's a matter for empirical investigation to establish which cognitive differences actually exist. As Horwich notes, one can't stipulate different cognitive processes for belief and acceptance without begging the question. But Horwich's own analysis grants us everything that's needed for making the distinction. Evidently, he accepts the proposition that we have sufficient cognitive resources for making both epistemic and pragmatic evaluations of theories. The precise manner in which these evaluations are performed needn't concern us. The fact that we can perform them is enough. Let's say that a program of action is *in accord* with a theory T if it's the rational program to adopt under the condition that T is true (given that our desires and our information are as they are). Now, we may commit ourselves to act in accord with T either because we consider T to have high epistemic value or because we consider T to have high pragmatic value. Our behavior might be the same in either case, but the two cases are different nonetheless, for the cognitive processes leading to the behavior are not the same. To be sure, we can't *stipulate* that these cognitive processes exist, but Horwich himself assures us that they do. Acting in accord with T because of its high epistemic value can be called *believing* T; acting in accord with T because of its high pragmatic value can be called *accepting* T. If it's objected that this isn't what belief and/or acceptance really are, we can once again take the recourse of introducing new terms like "epistemic belief" or "pragmatic acceptance". In any case, even if believers of T act exactly like accepters of T in every circumstance, constructive empiricism makes the coherent claim that they do so for different reasons. I do not myself endorse the view that belief and acceptance are the same type of propositional attitude. My point is that the coherence of constructive empiricism doesn't depend, as Horwich seems to suppose, on their being different attitudes. Their deployment of different evaluational criteria is enough.

In fairness to Horwich, it must be noted that he concedes that constructive empiricists might be able to save the coherence of their doctrine by reformulating it as a thesis about pragmatic versus epistemic grounds for theoretical belief. According to Horwich, when constructive empiricists are confronted with persuasive evidence for the identity of acceptance and belief, they can take recourse in the view

that "we ought indeed to believe theories given sufficient evidence, but . . . our rationale in such a case would be solely pragmatic and to no extent epistemic" (1991, 2).
Horwich proceeds to argue against the validity of this new version of constructive
empiricism, but he clearly concedes its coherence. This concession might seem to
be all that I'm presently arguing for. But there is a difference. Horwich regards this
reformulation of the constructive empiricist thesis as a *retreat to a weaker position* in
the face of a substantive point. Near the end of his article, he summarizes what he's
had to say as follows:

> I have tried to undermine two forms of instrumentalism: a stronger and a weaker
> version. According to the strong version, abductive inference is invalid and so it is
> legitimate to renounce theoretical belief. Against this I argued that there is no dif-
> ference between believing a theory and being disposed to use it; therefore the
> instrumentalist's proposal is incoherent. But this leaves open a weaker form of
> instrumentalism, namely, that since abductive inference is invalid, the justifica-
> tion for theoretical belief can be nothing more than pragmatic. (13)

My arguments aim to show that there's no important sense in which the reformulated doctrine is weaker than the original. The only substantive point of difference
between them concerns the proper interpretation of the folk-psychological concept
that goes by the name of "belief". This is not an issue in which antirealists have a
great deal of investment.

Finally, it should be noted that the entire issue is a minor skirmish in the battle
between realists and antirealists. If it should turn out, contra van Fraassen, that there
can be no coherent distinction between believing and accepting a theory, then a part
of the story that van Fraassen tells about our relation to theories will have to be given
up. But this part can easily be jettisoned without abandoning the ship of antirealism.
You don't have to "accept" theories to be an antirealist—you just have to refrain
from believing them. So, if theoretical acceptance is identical to theoretical belief,
antirealists can simply refrain from both.

The Theory-Observation Distinction I

Fodor's Distinction

What do Jerry Fordor and Bas van Fraassen, the archetypical sc ientific realist and his antirealist shadow, have in common? They're both defenders of the theory-observation distinction (van Fraassen 1980; Fodor 1984). It isn't surprising that a realist and an antirealist should agree about something, but it is curious that van Fraassen's and Fodor's defenses of the theory-observation distinction play diametrically opposite roles in their philosophical agendas. Van Fraassen needs it to support his antirealism; Fodor wants it in support of his realism. Van Fraassen needs the distinction to protect antirealism from the charge of incoherence. Antirealists wish to ascribe a more exalted epistemic status to the observational import of a theory than to its nonobservational claims. This position would be utterly untenable if the observational import couldn't be distinguished from the rest. And, in fact, denying the coherence of the theory-observation distinction has been a popular realist gambit in the debate about scientific realism (Maxwell 1962; Friedman 1982; Foss 1984; Creath 1985; Musgrave 1985).

Now, it might be claimed that realists also need the theory-observation distinction to state their thesis coherently — for how can we say that theoretical entities *do* exist, or that we're justified in believing in their existence, if there's no difference between the theoretical and the observational? This point is certainly applicable to some formulations of scientific realism. In fact, it applies to the definition that I gave in chapter 1. Yet the greater dependence of antirealism on the distinction is intuitively compelling. The lack of a theory-observation distinction might have consequences that realists would like to avoid (Fodor is about to acquaint us with a candidate for such a problematic consequence), but there can *be* no antirealism — not even a problematic antirealism — without a theory-observation distinction. The way out of this minor muddle is to say that scientific realists may either (1) accept the theory-observation distinction and say that some hypotheses of either type are believable, or (2) deny that the distinction is coherent and say that the realm of the believable includes some hypotheses belonging to the broader category that the theory-observation distinction tries but fails to dichotomize. As far as their debate with antirealists is concerned, the theory-observation distinction is optional for realists. They have everything to gain and nothing to lose by trying to make it unravel.

But antirealists aren't the only epistemic enemies of realism. Both realists and antirealists agree that there are *some* hypotheses and *some* logically and nomologically possible states of affairs such that we're *absolutely* warranted in believing the hypothesis if we find ourselves in the indicated state. This principle is denied by the *relativists*, who want to substitute warranted belief relative to a paradigm (or a culture, or an individual, etc.) for warranted belief *tout court*. Relativism is also a species of "antirealism", since it denies something that realists assert, but I reserve the label of "antirealism" for views like van Fraassen's that regard theoretical assertions as more problematic than observational claims. Note that there are issues wherein realists and relativists end up on the same side against the antirealists. In particular, realists and relativists agree that theoretical and observational hypotheses, if they can be distinguished at all, are in the same epistemic boat. They just differ as to the nature of the boat.

Anyway, Fodor thinks that a serviceable theory-observation distinction is needed to protect realism from the type of relativism that draws its inspiration from Thomas Kuhn. The Kuhnian relativist line is familiar: if there is no theory-neutral language with which to adjudicate our differences, we get incommensurability between supposedly competing theories; incommensurability brings in its train the irrationality of theory choice; and the irrationality of theory choice undermines the realist (and antirealist) claim that there can be scientific hypotheses that it's (absolutely) rational to believe. If this diagnosis is right, then realists have to choose their argumentative strategy with care. If they endorse the coherence of the theory-observation distinction, they protect themselves from the Kuhnian argument, but at the cost of giving up a devastating critique of antirealism. On the other hand, if they endorse the *inco*herence of the theory-observation distinction, it seems that they defeat the antirealists but lose to the relativists. Of course, to vanquish one philosophical school and to lose to another is to lose. Given these choices, the only viable option for realists is to try to show that the theory-observation *is* coherent, and to hope that the defeat of the antirealists will be realized in another arena. This is the course that Fodor takes.

I have three things to say about these issues. The first is that, contra Fodor, realists *don't* need a theory-observation distinction to protect themselves from the Kuhnian argument. If this is right, then they're free to go for a quick victory over the antirealists by blasting the distinction if they can. In a way, it's obvious that the lack of a theory-observation distinction does not, by itself, undermine realism. For suppose that there is no theory-observation distinction. Then realists can't appeal to observational hypotheses as the neutral ground upon which adherents to different paradigms may achieve a consensus. But this doesn't rule out the possibility that there's some *other* distinction, which may have nothing to do with theoreticity or observationality, but which nonetheless provides the requisite neutral ground. For instance, it doesn't rule out the philosophy of *sesquicentenarism*, which is the view that the first 150 mutually consistent hypotheses to be proposed must be recognized as true by adherents to any theory. If sesquicentenarism were true, then the relativist argument would fail even though the theory-observation distinction proved to be incoherent. On the other hand, the coherence of the sesquicentenarist's distinction, and even its epistemic relevance, would be of no use to antirealists—for it makes no connection to the empiricist scruples that underwrite their skepticism about theoretical entities. The neutral ground that seems to be needed to combat relativism doesn't need to have the observational character that antirealists require for their purpose. Sesquicentenarism, of course, isn't seriously in the running. In fact, in the debate between realists and Kuhnian relativists, it's been tacitly assumed that the observational is the only *candidate* for the requisite neutral ground. I give reasons to cast some doubt on this presumption in chapter 11. But until then, I accept it for the sake of the argument and show that realists *still* don't need a theory-observation distinction to ward off the Kuhnian argument. That is to say, realists don't need a neutral ground.

Second, I argue that neither of the two extant modes of making the theory-observation distinction—Fodor's and van Fraassen's—will do the job that antirealists require. If the analysis were to stop at this point, antirealism would be out of the picture, since there would be no satisfactory way of formulating what it claims. On the other hand, since realism doesn't need a theory-observation distinction, it would still be in the aforementioned picture along with relativism. My third point, however, is that there's a *third* distinction—one that partakes of some of the properties of both Fodor's and van Fraassen's distinctions—that (a) is arguably coherent, (b) arguably distinguishes two classes of hypotheses that differ in an epistemically significant manner, and (c) is arguably a theory-observation distinction. The viability of antirealism requires that all three of these propositions remain at least on the level of the arguable. If the third distinction is shown to be incoherent, for instance, then there doesn't remain any candidate distinction on the table with which antirealists can hope to formulate their creed. And even if the third distinction is coherent, antirealism is *still* out of the picture if it can be shown either that the distinction is irrelevant to epistemic concerns or that, as in the case of the sesquicentenarists' distinction, it isn't a theory-observation distinction.

What makes a distinction a theory-observation distinction? My main discussion (such as it is) of this topic is deferred until chapter 11. The topic briefly intrudes in a couple of places before then, however.

9.1 Fodor's Theory-Observation Distinction

Fodor discusses three arguments against the theory-observation distinction. Only one of his counterarguments, however, generates any philosophical heat. The first—the "ordinary-language" argument—has it that there is a theory-observation distinction, but that it's quite unsuited to playing the roles assigned to it in the grand epistemic debates in which it's been deployed. According to this argument, what counts as an observation depends on which issues an experiment is designed to settle and which assumptions are taken for granted as background knowledge. But, depending on our interests and the state of our science, *any* assumption might be taken for granted as background knowledge. Thus, any statement might, in appropriate circumstances, be granted the status of observation statement. Obviously, such a notion is of no help in combating the problem of incommensurability. There's nothing in this idea of observability that disallows the possibility that one scientist's observation statements are incommensurable with another scientist's observation statements. But this doesn't mean that some other way of making the distinction won't do the job:

> True, there is an epistemologically important distinction that it's reasonable to call "the" observation/inference distinction, and that is theory-relative. And, also true, it is this theory-relative distinction that scientists usually use the terms "observed" and "inferred" to mark. But that is quite compatible with there being another distinction, which it is also reasonable to call "the" observation/inference distinction, which is also of central significance to the epistemology of science, and which is *not* theory-relative. (Fodor 1984, 26)

Enough said.

Fodor next turns to the "meaning holism" argument against the "observation/inference distinction". The premise of this argument is the Quinean thesis that the meaning of any statement in a theory is determined by its relations to all the other theoretical statements. If this is so, then putative observation statements will have different meanings in the context of different theories. The conclusion is the same as the previous argument's: no common coin with which to purchase our way out of the incommensurability dilemma. Whereas the ordinary-language argument was too puny to warrant much of Fodor's attention, holism is too complex an issue for him to resolve in the compass of a journal article. Rather than refuting the argument from holism, Fodor contents himself with the observation that the truth of holism has not been established beyond all reasonable doubt. So, fans of theory-neutral language can at least hope for the best.

The argument that occupies most of Fodor's attention is the "psychological" argument, according to which the empirical results of psychological research teach us that there *is* no theory-neutral observation. Here is what the first Kuhnian relativist has to say:

> Surveying the rich experimental literature [of psychology] . . . makes one suspect that something like a paradigm is prerequisite to perception itself. What a man sees depends both upon what he looks at and also upon what his previous visual-conceptual experience has taught him to see. (Kuhn 1962, 113)

The psychology that teaches us this lesson is primarily the old "New Look" perceptual psychology of the 1950s. Bruner and his colleagues conducted a series of seminal experiments that purported to show that one's expectations—more generally, one's background theories—influence perception (Bruner 1957). From this general principle, it's supposed to follow that possessors of different theories will "see different things when they look from the same point in the same direction" (Kuhn 1962, 150). From the fact that scientists with different theories see different things, it follows that there is no such thing as theory-neutral observation, and from *this* it follows that there can be no theory-neutral observation *language*, for the simple reason that there's nothing for such a language to be about. The result is incommensurability for sure, and maybe even the irrationality of theory choice.

This line of reasoning is liable to several criticisms. Gilman (1992), for instance, argues that the New Look studies provide scant evidence for the hypothesis that one's theoretical equipment influences perception—for all their findings can just as well be explained by alternative and at least equally plausible hypotheses. One of these is the hypothesis that the subjects enjoyed a theory-neutral perceptual experience, but that when it came to reporting what they saw, they chose to correct their account of the event in light of their background theories. Gilman also makes the more general point that it's probably not a good idea to draw profound epistemological conclusions from the theory and research of so immature a science as psychology. Both these points are well taken. But Fodor pinpoints a more fundamental difficulty with the Kuhnian case for the nomological impossibility of theory-free observation. Let's grant that the New Lookers have established that one's background theory influences perception. Even so, it doesn't follow that scientists with different theories will always and inevitably see different things. We can accept the New Look hypothesis that cognition influences perception while still maintaining that there are *some* cognitive differences between scientists that make no perceptual differences. For instance, it may be that a knowledge of particle physics alters one's perception of particle tracks in cloud chambers, but that learning one more decimal place of π doesn't have any effect at all on perception. But then, even if the New Look hypothesis is true, it's possible for scientists who hold different theories to see the same thing when they look in the same direction—it just may be that the cognitive difference between them is one of those that doesn't make any perceptual difference. In fact, it's possible that *none* of the theoretical differences among scientists makes any difference to perception. To show the impossibility of theory-neutral observation, one would have to establish that *all* cognitive differences have an effect on perception—and this goes beyond what the New Look research program has established on even the most sanguine reading.[1]

In fact, Fodor claims that there's compelling empirical evidence for the proposition that perception *does* remain uninfluenced by many of our background beliefs. The most persuasive is the evidence for the persistence of perceptual illusions:

> The Müller-Lyer illusion is a *familiar* illusion; the news has pretty well gotten around by now. So, it's part of the "background theory" of anybody who lives in this culture and is at all into pop psychology that displays [of the Müller-Lyer illusion] are in fact misleading and that it always turns out, on measurement, that the center lines of the arrows are the same length. Query: *Why isn't perception penetrated by THAT piece of background theory?* . . . This sort of consideration doesn't make it seem at all as though perception is, as it's often said to be, saturated with cognition through and through. On the contrary, it suggests just the reverse: that how the world looks can be peculiarly unaffected by how one knows it to be. (1984, 34)

The upshot of this reflection on the persistence of illusions is that perceptual processes are, at least to some extent, impervious to top-down effects of cognition. In Fodor's terminology, perception is "informationally encapsulated": only a restricted range of information is capable of influencing the output of perceptual processes (1983, 64). This is, of course, a part of the broader theory that perceptual systems are modular.

> The intended epistemological moral of this story is clear: . . . if perceptual processes are modular, then, by definition, bodies of theories that are inaccessible to the modules *do not affect the way the perceiver sees the world*. Specifically, perceivers who differ profoundly in their background theories—scientists with quite dif-

ferent axes to grind, for example, might nevertheless see the world in *exactly* the same way, so long as the bodies of theory that they disagree about are inaccessible to their perceptual mechanisms. (1984, 38)

Moreover, the possibility of theory-neutral observation brings in its train the possibility of a theory-neutral observation *language*:

Suppose that perceptual mechanisms are modular and that the body of background theory accessible to processes of perceptual integration is therefore rigidly fixed. By hypothesis, only those properties of the distal stimulus count as observable which terms in the *accessible* background theory denote. The point is, no doubt, empirical, but I am willing to bet lots that "red" will prove to be observational by this criterion and that "proton" will not. This is, of course, just a way of betting that . . . physics doesn't belong to the accessible background. (1984, 38)

And thus ends the threat of Kuhnian incommensurability.
    Or does it?

9.2 Does Fodor's Distinction Do the Job That Fodor Wants?

Let us pause to appreciate what Fodor concedes. On his account, perception *is* determined in part by one's background information, and not just by the properties of the sensory array. He *has* to concede this on far more general grounds than the arguments of the New Lookers. He has to concede it because perception is obviously underdetermined by the properties of the sensory array: "as a matter of principle, any given pattern of proximal stimulation is compatible with a great variety of distal causes" (1984, 30). For example, one and the same pattern of proximal stimulation is compatible with the following distal causes: lines A and B are the same length and at the same distance from the observer; A is twice as long and twice as far away as B; A is three times as long and three times as far away as B; and so on. But then Fodor asks:

How is it possible that perception should ever manage to be *univocal* (to say nothing of *veridical*)? Why, that is, doesn't the world look to be many ways ambiguous, with one "reading" of the ambiguity corresponding to each distal layout that is compatible with the current sensory excitation? (31)

Fodor admits that there's only one answer to this question in the field:

Though perceptual analyses are underdetermined by sensory arrays, it does not follow that they are underdetermined *tout court*. For, perceptual analyses are constrained not just by the available sensory information, but also by such prior knowledge as the perceiver may bring to the task. What happens in perceptual processing, according to this account, is that sensory information is interpreted by reference to the perceiver's background theories, the latter serving, in effect, to rule out certain etiologies as implausible causal histories for the present sensory array. Only thus is sensory ambiguity resolved. (31)

It's important to understand that Fodor does not reject the thesis that background knowledge has a hand in shaping perception. He merely makes the point that perceptual processes are not penetrable by *any and all* background knowledge, and that this exemption establishes the possibility that some theoretical differences may be adjudicated by a common pool of perceptual experiences. It's true that perception itself is determined in part by our theories of the world, but if the theories in question are shared by all human beings, there continues to be a one-to-one relation between the sensory array and human perceptual experience. That is to say, the common coin needed for overcoming incommensurability is still available. Even if we admit that perceptual processes are penetrable by *some* beliefs acquired by training and experience, there's still the possibility of settling *some* theoretical disputes by observation.

But the fact remains that what we see depends on our background theories, whether endogenously specified or not. This has led Churchland to make the following objection:

> Let us suppose . . . that our perceptual modules . . . embody a systematic set of . . . assumptions about the world, whose influence on perceptual processing is unaffected by further or contrary information. . . . This may be a recipe for a certain limited *consensus* among human perceivers, but it is hardly a recipe for theoretical *neutrality*. . . . What we have is a universal dogmatism, not an innocent Eden of objectivity. (1988, 169–170)

In his reply to this objection, Fodor concedes that reliance on the perceptual information supplied by informationally encapsulated perceptual modules constitutes a perceptual bias, and that it makes perception "inferential" (Fodor 1988, 189). His point is that this bias is compatible with the negation of the incommensurability thesis: to the extent that the bias is universal, it will not be an impediment to the resolution of theoretical disputes on the basis of scientists' perceptual experiences. Fodor and Churchland are arguing at cross-purposes here. Churchland's point is that Fodor's distinction doesn't give us any reason to suppose that observation statements are on a firmer epistemic footing than theoretical statements. This might mean that *anti*realists can't use the distinction for *their* purposes—I consider whether this is so in section 9.3. But Churchland's point isn't one that a *realist* needs to worry about. Fodor isn't looking for a notion of observationality that underwrites our granting epistemic privilege to observation statements. He's looking for a notion that wards off the incommensurability argument. And for that purpose, anything that produces "consensus" will do. If it could be shown that all scientists, regardless of theoretical affiliation, are sesquicentenarists, that would do the job just as well.

So, Fodor's way of making the distinction will do the job that Fodor wants it to do—*if* the modularity theory is true. But Churchland also challenges its truth. He claims that it's obvious that acquired expertise alters the perceptions of the expert. For example, someone musically sophisticated "perceives, in any composition whether great or mundane, a structure, development and rationale that is lost on the untrained ear" (1988, 179). This point has long been a mainstay of the Kuhnian line:

> Seeing water droplets or a needle against a numerical scale is a primitive perceptual experience for the man unacquainted with cloud chambers and ammeters. It thus requires contemplation, analysis, and interpretation (or else the intervention of external authority) before conclusions can be reached about electrons or cur-

rents. But the position of the man who has learned about these instruments and had much exemplary experience with them is very different, and there are corresponding differences in the way he processes the stimuli that reach him from them. Regarding the vapor in his breath on a cold winter afternoon, his sensation may be the same as that of a layman, but viewing a cloud chamber he sees (here *literally*) not droplets but the tracks of electron, alpha particles, and so on. (Kuhn 1962, 97, emphasis added)

Fodor's reply is that Churchland and Kuhn merely beg the question of whether this effect is, in fact, perceptual:

What Churchland has to show is . . . that *perceptual* capacities are altered by learning musical theory (as opposed to the truism that learning musical theory alters what you know about music). (1988, 195)

Fodor seems to assume that the burden of proof is on the opposition: "You don't refuse modularity theory by the unsupported assumption that it is contrary to fact" (195). Fair enough. But you don't *support* the modularity thesis by the mere assertion that the phenomenon is *not* contrary to the theory, either. We start with the incontestable datum that physicists *talk* about seeing electron tracks in cloud chambers. The question is how this sort of talk is to be understood. Presumably, there's one way of understanding it that supports Churchland's view and another that supports Fodor's. It seems to me that burden-of-proof considerations are symmetrically distributed in this case.

Actually, it's not immediately clear just what distinguishes the Fodorean and the Churchlandian interpretations. Fodor has admitted that the deliverances of the perceptual module are the result of an inferential process, so the issue isn't whether Kuhn's physicist infers his way to electron tracks or sees them "directly". Nor can the difference be between conscious and unconscious inferential processes—for Fodor would surely be willing to concede that the physicist's inferences to electron tracks can become so habitual that they're automatized and fall below the threshold of consciousness. Does the putative difference between the physicist and the layman reside in the qualitative content of their experiences? Until this point, qualia have played no role in Fodor's arguments. The arguments have relied entirely on a functional notion of perception. As far as I can tell, nothing in Fodor's analysis would be changed by the assumption that observations—that is, the outputs of perceptual modules—were utterly bereft of qualia. If Fodor has a story to tell about the theory-observation distinction that involves qualia, he has not shared it with us.

The difference between seeing that there are electron tracks and merely making the quick unconscious inference that there are electron tracks can only be construed as a difference between two relatively complicated theoretical stories. Fodor's version runs something like this. Patterns of sensory stimulation are inputs into a perceptual module P whose mode of operation is innately specified, and whose outputs are representations of the scene before us; these outputs are in turn inputs into the central system C that is stocked with various concepts and theories; and belief fixation is the result of applying the notions in C to the input from P.[2] When standing in front of a cloud chamber, the output of P for physicist and layman alike is that there are trails of droplets—this is the sense in which the physicist and the layman see the same thing. But when these identical outputs are processed by the physicist's and layman's central systems, the result, by virtue of their Cs being stocked with different concepts and theories, is the fixation and subsequent report of different beliefs. Let's call this the *Fodorean* model of perception.

Churchland's view is that this theoretical story is false in some important respect: either (1) C has collateral inputs from the sensory transducers that bypass P, or (2) C has a top-down influence on the operation of P, or (3) there are no Ps at all—that is, there is no principled distinction between perception and cognition. Any of these three scenarios allows us to say that there is, or may be, no systematic difference between the layman's seeing of droplets and the physicist's seeing of electron tracks. There are interesting and potentially testable consequences that distinguish each of these three models from the other two. Most of these differences are irrelevant to our present concerns, however. In particular, there's no relevant difference between (2) and (3). So I'll let my discussion of (3) stand for both. I'll call (3) the *Kuhnian* model of perception, and (1) the *mixed* model.

Both the Kuhnian and the mixed models incorporate the assumption that the sensory transducers feed their output directly into the central system. On this view, our acquired theories can contribute to the operations of *assembled* (i.e., not hard-wired and not innate) perceptual systems (henceforth, APs) whose output is going to vary depending on which theories they embody. In the Kuhnian model, the outputs of APs comprise all the perceptual information available to C, the central system responsible for belief fixation. In the mixed model, the outputs of both our endogenous P and our acquired APs are available as inputs to C. Maybe this architectural situation produces Gestalt switches in us—sometimes we see droplets, sometimes electron tracks—or maybe both outputs are in some sense "available" at the same time. These are the kinds of theoretical details that can safely be ignored at present. In any case, it's clear that Fodor wants to reject this notion of assembled perceptual systems that is common to both the Kuhnian and the mixed models. And so does Gilman:

> Recall the Kuhnian example of someone who "literally" sees the tracks of electrons, alpha particles, and so forth as opposed to droplets. Or the well-known example of the scientist who always sees a pendulum as a pendulum. Surely none would deny that such people can explain what they are noticing to people not conversant with the relevant concepts (or to people who are sceptical about the evidence) by pointing to the droplets as droplets. Surely a scientist is capable of turning to a scientist's child and saying: "look there, you see how that weight is swinging at the end of the line . . . " when the child says that he does not know what a pendulum is. (1992, 307)

I interpret this as a rejection of the existence of APs for the following reason. Gilman's argument is that the deliverances of the physicist's perceptual system can be represented in the format used by P, the innately specified perceptual module that is common to all humanity. But if the outputs of the two systems are intertranslatable, then there is no theoretical rationale for making the distinction between the systems in the first place. If you can say what you see when you see electron tracks in Perceptualese (the code in which P formats its outputs), then you don't need anything more than P to see it. So, the fact, if it is a fact, that the physicist can tell the layman what he sees when he sees particle tracks shows that they can see the same thing.

I have two things to say about this argument. The first, and lesser, point is that it's inconclusive, especially for someone who holds Fodorean views about the nature of central processes. Let's grant Gilman's premise that a physicist can teach a layman to discriminate what he discriminates by pointing to features that the layman is already able to perceive. That this should be the case isn't surprising, since the physicist himself had to learn to make the same discriminations by listening to the explanations of *his* teacher. What is perhaps surprising, if it's true, is that we can learn to make the relevant discriminations so quickly. One might have thought that it would take a least a semester of graduate study in particle physics. But whether it takes a few moments or a few years, we can very well grant that having our attention called to Perceptualese features of the scene causes us to see what physicists see *without* granting that what physicists see can be described in Perceptualese. If one grants that perceptual systems can be assembled, there has to be *some* causal story about how the assembling gets done. Maybe we come to see particle tracks by taking a psychotropic drug—or maybe it's having certain Perceptualese features pointed out to us that does the job. Certainly there's no a priori reason to exclude the second possibility. For someone who holds Fodor's views about central processes, it shouldn't even be very surprising if the whole procedure takes only a few moments. Fodor famously argues that all the concepts available to C are themselves endogenously determined: either they're in residence right from the start, or their *seed* is in residence, waiting to be triggered by the appropriate environmental event (Fodor 1981). On this account, the layman already *has* the concept of an electron, at least in potential form. So, even if having the concept of an electron is a necessary precondition for seeing the same thing as a physicist, the layman may need no more than to have somebody pull the electron-trigger. In sum, Gilman's observation that physicists can point out pendulums and electron tracks to laymen is not very strong evidence against the existence of APs.

The second and more important point about Gilman's argument is that there's no reason why a Fodorean realist should *care* whether there are APs. For the existence of APs is compatible with the mixed model, according to which we enjoy the benefits of an endogenously specified P in *addition* to assembled APs. As long as Fodor is granted his P, he's assured of a pool of common human perceptions, and as long as *some* of the implications of our theories can be described in Perceptualese, we can hope to settle (some of) our theoretical differences on the basis of the observational evidence. The fact that our theories may *also* have observational implications that are incommensurable with each other doesn't threaten the objectivity of science—it just adds a little spice to the proceedings. The sequence of ideas leading from Kuhn's analysis, through Fodor's critique and Churchland's countercritique, to my present point can be represented as a growing awareness of the importance of missing quantifiers. In the beginning, Kuhn asserts that there's an incommensurability problem because of the fact that our theories influence what we observe. Fodor makes the point that there's no problem unless *all* our theories influence what we observe. Churchland claims that all our theories, or indefinitely many of them, *do* influence perception, and Fodor and Gilman try to refute this claim, or at least to show it to be unsupported. My point is that even if Churchland is right, no *serious* problem about incommensurability ensues. For there to be a serious problem, it would have to be shown that our theories influence *all* of our perceptions.[3]

Now let's consider the implications of the Kuhnian model, the one that says that there is no P. If this model is correct, the only way we can see the world is via the outputs of APs. What is the character of perceptual experience prior to the assembly of the first AP? Here's what the eponymous proponent of the Kuhnian model thinks:

> What a man sees depends both upon what he looks at and also upon what his pre-vious visual-conceptual experience has taught him to see. In the absence of such training, there can only be, in William James's phrase, "a bloomin' buzzin' confu-sion". (Kuhn, 1962, 113)

And what's the import of this theoretical change on the incommensurability prob-lem? Well, it allows for the possibility that different groups of individuals, by virtue of having different APs, will not be able to settle their theoretical differences by empirical means. They "live in different worlds". This is, of course, the problem that we started with. So, it seems that the existence of P, with or without collateral APs, is a prerequisite for Fodor's defense against the relativists.

However, a closer look reveals that the epistemic situation vis-à-vis incommen-surability is essentially the same in the Kuhnian model as in the mixed model. Sup-pose scientist S1 has the assembled perceptual system AP1 and S2 doesn't. Then S2 can't confirm or disconfirm any of S1's theoretical claims—*in his present state*. But if S1 has managed to assemble AP1, there's no reason why S2 can't assemble it as well. In fact, there's no reason why *anyone* can't assemble AP1. If everyone *did* assemble AP1, then the epistemic situation would be identical to the situation that results from our all sharing the endogenously specified perceptual module P. In particular, we would all be able to confirm or disconfirm the claims of theories that have conse-quences that can be put in the format of AP1. To be sure, everyone might *not* as-semble AP1. Maybe the assembly of AP1 requires the equivalent of several years of graduate study in physics. Maybe it requires our living for years among the Azande. So, some people are going to be unable, because of their sloth, to confirm or dis-confirm the theories that possessors of AP1 can evaluate.

But this has no greater epistemological significance than the fact that some pos-sessors of P, the endogenous perceptual module, may systematically fail to *utilize* the capacities of P in their theoretical deliberations. Suppose once again that we all share the same endogenous P. Suppose also that scientist S1 has the assembled per-ceptual system AP1 and that S2 has the system AP2. Now consider two theories T1 and T2 such that both T1 and T2 have Perceptualese consequences—that is, the deliverances of P may confirm or disconfirm T1 and T2. In fact, suppose that T1 and T2 make conflicting Perceptualese predictions, so that it would in principle be pos-sible to choose between them on the basis of P's outputs alone. In addition, how-ever, T1 has implications for the output that one expects to obtain from AP1, and T2 has implications for the output of AP2. Finally, suppose that S1 and S2 both have the bad habit of being utterly inattentive to the deliverances of P. Looking at a cloud chamber, S1 might see particle tracks and simply *not notice* that there are droplets.

In that case, a state of de facto incommensuration obtains between S1 and S2. It's true that they *could* take steps to resolve their differences, but this is precluded by their habits of inattention to P. I claim that this situation does not differ in any epistemically relevant manner from the situation wherein there is no P, and S1 and S2 have only the assembled and incommensurable paradigms AP1 and AP2 to rely on for observational information. The incommensurability between them is just as remediable in this case as in the preceding one: S1 and S2 could simply assemble each others' AP. The fact that some people — or even most people — may not choose to do so has no greater significance than the fact that some people may not choose to attend to the deliverances of their endogenously specified perceptual modules. If you don't pay attention to what the world looks like, then you can't rationally make the same theoretical choices as people who *do* pay attention to what the world looks like. And if you don't learn physics, or live with the Azande, then, once again, you're barred from making certain theoretical choices that are within the range of those who are more energetic or dedicated than you are. What's the difference? Both deficiencies are equally problematic; both are remediable. Their epistemological consequences should be the same.

Here are four potential objections to the foregoing line, in order of increasing severity. First problem: it's not the case that everyone can assemble any perceptual system, for the acquisition of some APs requires extraordinary qualities of intellect or peculiarities of character. Therefore, there is no universally shared pool of perceptions without an endogenously specified P. Reply: there isn't going to be a truly *universal* pool of experiences *with* endogenously specified Ps, either. Even if all visual perceptions are the outputs of an endogenously specified visual system, there are going to be some people who won't be able to obtain these perceptions, regardless of how earnestly they try. We call them the blind. We don't have the option of insisting that any class of perceptions has to be universally available. Our epistemology has to take into account the fact that some people suffer from perceptual deficiencies that place restrictions on their degree of participation in epistemic affairs. On the model we're considering, the inability of some people to assemble some APs, no matter how hard they try, is, among other things, a perceptual deficiency.

Problem 2: APs (strictly, the theories embodied by APs) are self-validating — there's no possibility of disconfirming them by observational means, since they determine what our observations are going to be. Hence, there's no accounting for rational theory change if all perceptions are funneled through APs. Reply: there's no less reason to say the same about P. Apparently, Fodor believes that the theoretical assumptions embodied by P are compatible with a range of theories, some of which may be direct competitors with one another. On Fodor's view, this is the range of theories concerning which we can rationally resolve our differences by observational means. Why should it be any different with assembled perceptual systems than it is with endogenously specified systems?

Problem 3: you can't simultaneously have two APs; therefore, if S1, who has paradigm AP1, grows paradigm AP2, S1 simultaneously loses AP1 and, with it, the ability to assess the theories whose implications are in the format of AP1. First reply: so what? Second reply: who says you can't have two APs at once? Certainly, there's no architectural problem in possessing two disparate perceptual systems that are connected at one end to the central system and at the other end to one and the same sensory transducer. In fact, orthodox modularity theory posits just such parallel connections involving the endogenously specified language comprehension module and the endogenously specified acoustic module. Both of them are hooked up to the same auditory transducer, but one of them outputs token sentences while the other puts out the sound of wailing saxophones and fingernails against the blackboard. Why shouldn't two assembled systems be hooked up in the same fashion?

Problem 4: to "have" a perceptual system entails that you have certain *beliefs*— namely, those that are embodied by the system. But you can't choose to believe something just because you want to have access to a certain set of perceptions. Also, the fact that belief is a necessary condition for perceptual-system-having is a reason why you can't simultaneously have two systems that embody conflicting theories. Let's dispose of the latter corollary first. If the theories embodied by two perceptual systems conflict in the sense that one of them entails X and the other entails not-X, then there's no incommensurability problem between them in the first place (since X is describable in both). And if they don't conflict in this sense, then there's no logical impropriety in believing them both. After all, don't we generally believe in the deliverances of both the acoustic and the language comprehension modules? As for the main point: I admit that to the extent that what we perceive is shaped by what we *believe*, we face a fearsome pack of puzzles and paradoxes. I've made an initial foray into this conceptual minefield in another essay (Kukla 1994). But it isn't at all obvious that assembling a perceptual system requires us to believe in the theory that it embodies. I don't have to believe that what I see is really a duck when I see the duck-rabbit as a duck; I don't even have to believe that ducks *exist*. By the same token, I bet that I could learn to see the phenomena relating to combustion phlogistically without believing the phlogiston theory. I just need to have mastered the conceptual machinery of the theory—I don't have to be committed to it. Similarly, if belief in elementary particle theory were a precondition for seeing particle tracks in cloud chambers, then you wouldn't be able to see them if you were a *instrumentalist*, no matter how steeped you were in the theory. And if a physicist were persuaded by van Fraassian arguments to *become* an instrumentalist, his perceptions would automatically revert to the layman's. All this strikes me as utterly implausible. I take up this topic again in chapter 11. At any rate, it's premature to worry about the possibility that you can't assemble an AP without believing its constitutive assumptions.

It's time to recapitulate, draw a conclusion, and move on. The coherence of Fodor's theory-observation distinction depends on the truth of the Fodorean model of perception, or at least of the mixed model. I've expressed no opinion as to the truth or falsehood of this disjunction. Actually, I think that the hypothesis that we have endogenously specified perceptual modules has a lot going for it. But the point of my discussion is that the truth or falsehood of this hypothesis, hence also the coherence or incoherence of Fodor's theory-observation distinction, makes no difference to the issues relating to incommensurability and the rationality of theory choice. The epistemological situation is essentially the same whether we perceive the world exclusively through endogenously specified modules, or exclusively through assembled perceptual systems, or through both. Realists don't *need* Fodor's distinction in their struggles against the Kuhnian relativists, and it wouldn't help them if they had it.

9.3 Does Fodor's Distinction Do the Job That van Fraassen Wants?

We've seen that Fodor's theory-observation distinction isn't needed for the job that Fodor fashioned it to perform. But maybe the antirealists can get some use out of it. What antirealists need is a distinction that has a prima facie chance of sustaining their thesis that the observational parts of our theories should be granted certain epistemic privileges that are to be denied to the nonobservational parts. It isn't necessary that the antirealists' thesis be an obvious consequence of their distinction — that's too much to ask. What they're looking for — or at least, the portion of what they're looking for that this chapter is concerned with — is a rebuttal to the frequent realist charge that the very distinction between the observational and the nonobservational is incoherent or otherwise unusable by antirealists. Assuming that antirealists can navigate successfully past this fundamental obstacle, they can then hope to make the rest of their case with dialectical weapons such as the argument from the underdetermination of theory by observational data. The question here is whether there's a way of rendering the theory-observation distinction that allows them to graduate to the second round of argumentation.

Objections to antirealist distinctions between the observable and the nonobservable can be divided into the *devastating* and the merely *troublesome*. Devastating objections are those which show that the distinction is definitely unusable in second-round arguments about epistemic privilege. A distinction is devastated if it's shown to entail (1) that one and the same thing is both observable and not observable, or (2) that all of a theory's import is observational, or that it's all nonobservational, or (3) that antirealism is false. Distinctions that can't be devastated are *compatible* with antirealism, but they don't necessarily *support* it. Distinctions are *troubled* if there's nothing in one's present system of beliefs that can bridge the gulf between making the distinction and concluding that antirealism is the best epistemic policy, or if there is *something* in our present system of beliefs that *blocks* the move from the distinction to antirealism. If a distinction is troubled, then antirealists are required either to *change* their belief system (add a new belief or throw out an old one), or to look for a new distinction, or to give up their antirealism. Trouble happens when it's shown that antirealists can't make a case for their antirealism with the cognitive chips on the table. It's not just a matter of not having thought of an argument to do the job. The accusation that a distinction is troubled amounts to the claim that there isn't an argument to be found unless beliefs are added or subtracted. Naturally, the severity of the trouble varies with the centrality of the beliefs that need to be added or subtracted. However badly troubled a distinction may be, however, the antirealist may still allow herself to hope that some remedial work will make everything turn out well in the end.

What's the status of Fodor's distinction, vis-à-vis antirealism? Before talking about his *real* distinction, let's look at what he sometimes pretends his distinction to be.

Fodor routinely refers to his distinction as the "observation/inference" distinction. Here's how Granny, Fodor's traditionalist alter ego, puts it:

> "There are," Granny says, "two quite different routes to the fixation of belief. There is, on the one hand, belief fixation directly consequent upon the activation of the senses (belief fixation 'by observation', as I shall say for short) and there is belief fixation via inference from beliefs previously held ('theoretical' inference, as I shall say for short)." (1984, 23)[4]

The good thing about this distinction, from the antirealist's point of view, is that there's a prima facie case for its epistemic relevance. "Inferential" beliefs must ultimately be based on "observational" beliefs, on pain of infinite regress. Hence, the former inherit whatever epistemic risks the latter are afflicted with, and add these to the risks that accrue to inference-making. Thus, the inferential must be burdened with more epistemic risks in toto than the observational. So, it seems that the observation-versus-inference distinction is good enough to propel the antirealist to the second stage of argumentation. The problem, of course, is that Fodor's own model of perception renders this particular distinction unusable. For Fodor, observation (i.e., perception) is itself the result of an inferential process. If the Fodorean model is correct, all beliefs are inferential. It may be true that perceptual beliefs involve *fewer* inferences than theoretical beliefs—I consider that possibility below. But for those who subscribe to Fodor's model, to distinguish between (1) beliefs that are based on *some* inferences and (2) beliefs that are based on *no* inferences is not to distinguish among our beliefs at all. This means that Granny's distinction is *troubled*: if Granny wants to keep it, she's going to have to give up the Fodorean model. In fact, Granny's distinction can't be used by adherents to the Kuhnian or mixed model, either, since perception is inferential in these as well. If she wants to keep it, she'll have to devise a theory of perception along entirely different lines. (Of course, being a realist, Granny doesn't *need* to keep the distinction.)

What about Fodor's real distinction between the output of endogenously specified perceptual modules (P) and the output of the central system (C)? To be sure, the representation that travels from P into C is also a result of inferential processes, just like the representation that goes from C into the belief bin. But, one might argue, the final output from C will have gone through an *additional* inferential process. Hence, the situation is akin to that which obtains when we compare inferential with noninferential beliefs. Whatever risk of error there may be in converting the sensory array into a perception, the risk can only be compounded when we take the additional step of converting the perception to a centrally inferred belief. Granny's distinction between the inferential and the noninferential may be incompatible with the facts about perception, but the distinction between the more inferential output from C and the less inferential output from P may be able to stand in for it in the arguments of the antirealists. In fact, it can't. It's true that cognition is more inferential than perception in the Fodorean model. But that's not enough of a result to allow the antirealists to graduate to the second round of argumentation. The problem isn't that antirealists can't get by on a difference of degree—at least that's not the problem that I want to press. The problem is that antirealists need more than a dis-

tinction between *perceptions* and centrally inferred beliefs. What they need is a distinction between two categories of *beliefs*, such that one type has a prima facie greater
chance of getting into epistemic trouble than the other. To be sure, if we say that the
output from P is already propositional, then it is possible to make a direct comparison between its epistemic merits and those of the beliefs put out by C. But that's neither
here nor there. Antirealists don't want to say that the *perception* that the cat is on the
mat is epistemically more secure than the belief that electrons are coursing through
a cathode ray tube — they want to say that the *belief* that the cat is on the mat is more
secure than beliefs about electrons. And the output of P isn't a belief. It doesn't *become* a belief until it passes through C.

Perhaps antirealists can make a distinction between "perceptual beliefs" and
"nonperceptual beliefs", where the former involve C's direct endorsement, as it were,
of the output it receives from P: we perceive that the book is on the table, and thereby
judge that the book is on the table. The idea here is that perceptual beliefs incur less
epistemic risk than nonperceptual beliefs because the latter involve an extra layer of
inference, while the former pass right through the central system without any additional processing. This suggestion won't work for two reasons.

First, supposing that C does allow the input from P to pass right through to the
belief bin, this process of passing right through is no less inferential than if the input
from P were drastically altered on its way to the belief bin. The inference is from
"the output of P is that X" to "X". This inference may be simpler in form than that
which our so-called nonperceptual beliefs go through, but this greater simplicity
doesn't *by itself* provide any reason to posit epistemic differences. After all, the inference from the perceptual input "the output of P is that X" to the belief "*not*-X" is also
simpler than the inferential transformations that most of our nonperceptual beliefs
go through, but we're not tempted on that account to suppose that it's likelier to be
valid. The fact that perceptual beliefs go through less radical transformations in C
than nonperceptual beliefs isn't by itself a reason to suppose that they have greater
epistemic merit.

Second, it isn't true that the only transformation applied by C to its inputs from
P is to change "the output of P is that X" into "X". If it were, then we would believe
that sticks bend when they're dipped in water and that one of the lines in the Müller-
Lyer illusion is longer than the other. As Fodor emphasizes in a number of places,
the perceptual outputs of P are routinely corrected by background knowledge before
they're converted into beliefs. So even when a perceptual output does pass through
C and exit into the belief bin in its original form, the free passage is granted on the
basis of *special consideration* by C. We don't automatically believe everything we
see. But then, when we *do* believe what we see, this belief is as dependent on a second layer of central-system inference-making as are our nonperceptual beliefs.

Fodor's distinction between beliefs that correspond to the deliverances of P and
beliefs that *don't* correspond to such deliverances is logically *compatible* with antirealism — there's no reason why antirealists shouldn't make this distinction if it strikes
their fancy to do so. But the problem for antirealists is that there isn't even a prima
facie reason to suppose that it has any connection to our epistemic concerns. We
could put the matter this way: in relation to the antirealists' philosophical agenda,
the Fodorean distinction between perceptual and nonperceptual beliefs is an *arbi-*

trary distinction. We will see this charge of arbitrariness again when we talk about van Fraassen's theory-observation distinction. What, exactly, should its impact on antirealists be? There's no doubt that it's a difficulty they should worry about. But the difficulty is once again merely troublesome rather than devastating. The charge of arbitrariness is the charge that antirealists currently have no resources for connecting up their distinction to their epistemological hypothesis. This leaves it open that they might yet be able to establish their hypothesis on the basis of the selfsame distinction. All they have to do is to garner additional epistemic resources. Of course, this rescue operation may not succeed. But it's impossible for their realist opponents to establish that an arbitrary distinction is *terminally* arbitrary—that is, that there's *nothing* that can be added to the antirealists' system of beliefs that would connect their distinction with their epistemology. For any distinction between two classes of beliefs, however arbitrarily drawn, there are always going to be auxiliary hypotheses that make that distinction epistemically significant. Even the sesquicentenarist's distinction between the first 150 beliefs and the rest would be epistemically significant under the auxiliary hypothesis that God inspires us with the truth the first 150 times we adopt a belief, just to get us off to a good start, and then delivers us to our own imperfect cognitive devices.

Moreover, even if it were agreed that a given distinction *is* irremediably arbitrary, it wouldn't immediately follow that it's inadequate to underwrite antirealism. For it's possible to combine antirealism with *epistemological* auxiliaries that tell us that it's sometimes rational to believe a hypothesis even though the choice between it and its negation is arbitrary. For example, consider the following case, which I discuss more fully elsewhere (Kukla 1994). Suppose that a particular hypothesis H is a self-fulfilling prophecy in the strong sense that one's believing H causes H to be true, and one's *not* believing H causes H to be false. It's important for the example that not only active belief in the negation of H, but also having no opinion about H (in sum, *non*belief in H) is sufficient for H's being false. Suppose now that you fully understand the strongly self-fulfilling nature of this hypothesis. In that case, there's neither more nor less reason for you to believe H than to believe not-H. Whatever you choose to do, you will not have acquired any false beliefs. The choice is arbitrary. Yet it isn't a mistake to choose. In fact, in this case, making a choice between believing H and believing not-H is *mandatory*: logic compels you either to believe H or *not* to believe H, but if you opt for nonbelief, then you know that H will turn out to be false; hence, suspending judgment is not an option. Thus, the fact that a belief is irremediably arbitrary doesn't, by itself, establish that it's irrational to adopt it. This is very close to the line that van Fraassen himself takes about antirealism in his *Laws and Symmetry*.

In sum, the fact that Fodor's theory-observation distinction is arbitrary regarding antirealism is a troublesome criticism, but it's not devastating. The charge of arbitrariness shows that antirealists owe us some further explanation—their case isn't complete—but it doesn't yet show that they're *wrong*. Still, it's nice to avoid trouble if you can. The troublesomeness of Fodor's distinction gives antirealists a philosophical motive for looking into alternative ways of making the distinction that might be trouble-free. There's also another reason for antirealists to look beyond Fodor's distinction: the distinction is incompatible with the Kuhnian model of perception, which posits the *absence* of endogenously specified perceptual modules. Whether this is a problem for proponents of Kuhn's model or users of Fodor's distinction is a moot point. But one thing is certain: if you're a believer in the Kuhnian model of perception and an antirealist, then you can't just throw Fodor's theory-observation distinction out the window. It's incumbent upon you to come up with another way of describing how your epistemically privileged class of hypotheses differs from the rest.

Which brings us to van Fraassen's theory-observation distinction.

The Theory-Observation Distinction II

Van Fraassen's Distinction

V an Fraassen never expected any aid or comfort from Fodor-style distinctions. His discussion *begins* with the assumption that the Kuhnian model is correct:

> All our language is thoroughly theory-infected. If we could cleanse our language of theory-laden terms, beginning with the recently introduced ones like "VHF receiver", continuing through "mass" and "impulse" to "element" and so on into the prehistory of language formation, we would end up with nothing useful. The way we talk, and scientists talk, is guided by the pictures provided by previously accepted theories. This is true also . . . of experimental reports. (1980, 14)

In this passage, van Fraassen doesn't explicitly subscribe to or repudiate any psychological model of perception. His thesis is about *language*. In particular, he denies that there can be a theory-neutral language. The connection between this linguistic thesis and psychological models of perception is pretty immediate, however. If Fodor's model of perception is true, then we can draw a distinction between the sentences of our public language that can be translated into Perceptualese (the output format of the perceptual module) and those that can't. And, by modus tollens, if "all our language is thoroughly theory-infected", then there must not be any perceptual module that delivers an output prior to the act of theoretical interpretation.

Van Fraassen's idea is to make the observational-nonobservational distinction in terms of *entities* instead of languages. Our scientific theories tell us, in their own theory-laden language, that certain entities or events impinge on our sensory transducers and that others don't. For example, science tells us that some complicated systems of elementary particles are of the right size and configuration for reflecting light in the portion of the spectrum to which our retina is sensitive. These systems are visually observable objects. On the other hand, isolated elementary particles are not observable. Antirealism can be understood as the thesis that beliefs that purport to be about observable objects or events have certain epistemic privileges that are denied beliefs that purport to be about unobservables. Van Fraassen's distinction, if it's coherent, is certainly compatible with the Kuhnian model. It's also compatible with the Fodorean and mixed models. In fact, it seems to be compatible with any psychological theory that posits the existence of sensory systems. But is it coherent? And if it is, does it have any serious troubles? Let's look at the arguments that have been leveled against it.

10.1 The Continuity Argument

The locus classicus for arguments against the theory-observation distinction is Grover Maxwell's (1962) "Ontological Status of Theoretical Entities". In this essay, Maxwell offers three arguments against the distinction. He contends

> that the line between the observable and the unobservable is diffuse, that it shifts from one scientific problem to another, and that it is constantly being pushed toward the "unobservable" end of the spectrum as we develop better means of observation. (13)

In 1962, the main proponents of antirealism were still the logical positivists, who conceived of the theory-observation distinction as a difference between two languages. As a consequence, Maxwell's arguments are directed to the linguistic distinction. But all three arguments apply just as well, mutatis mutandis, to the distinction between observable versus unobservable entities. The second argument—the charge that the distinction "shifts from one scientific problem to another"—is essentially the same as what Fodor calls the "ordinary-language" argument. Let's look at the other two.

Maxwell's first argument is that the distinction between observable and unobservable is a matter of degree rather than a dichotomy. He presents several graded series of events to persuade us of this thesis. The most persuasive, I think, is the series that begins with hydrogen molecules and ends with enormous lumps of plastic:

> Contemporary valency theory tells us that there is a virtually continuous transition from very small molecules (such as those of hydrogen) through "medium-sized" ones (such as those of the fatty acids, polypeptides, proteins, and viruses) to extremely large ones (such as crystals of the salts, diamonds, and lumps of polymeric plastic). The molecules in the last-mentioned group are macro, "directly observable" physical objects but are, nevertheless, genuine, single molecules; on the other hand, those in the first-mentioned group have the same perplexing properties as subatomic particles (de Broglie waves, Heisenberg indeterminacy, etc.). Are we to say that a large protein molecule (e.g., a virus) which can be "seen" only with an electron microscope is a little less real or exists to somewhat less an extent than does a molecule of a polymer which can be seen with an optical microscope? And does a hydrogen molecule partake of only an infinitesimal portion of existence or reality? Although there certainly is a continuous transition from observability to unobservability, any talk of such a continuity from full-blown existence to nonexistence is, clearly, nonsense. (1962, 9)

The argument, in brief, is that observability-unobservability is a continuum, whereas existence-nonexistence is a dichotomy; therefore, whether or not something exists can't be determined by its observational status.

The first and obvious antirealist retort to this argument is that it isn't necessary for antirealists to claim that only the observable *exists*. Van Fraassen certainly doesn't claim this. What van Fraassen and other *epistemic* antirealists want to say is that it is only information about the observable properties of observable things that is *believable*—and believability surely does come in degrees. So why can't antirealists say that, ceteris paribus, a claim is more believable to the extent that it deals with entities and events on the observable end of the continuum? Foss has noted that this "Bayesian" solution has the shortcoming that it would "dramatically reduce the difference between constructive empiricism and realism":

> If the constructive empiricist embraces the "Bayesian" solution . . . , then when he accepts a theory he will have various degrees of belief that each of the various theses of the theory is true. This position does not amount to being "agnostic about the existence of the unobservable aspects of the world described by science." . . . For the Bayesian sort of constructive empiricist does not suspend belief, but has quite definite degrees of belief about each scientific thesis. (1984, 85–86)

An unstated premise of this argument is that, observability being a matter of degree, there are no entities that are *totally* bereft of this quality—for if there were, then constructive empiricists could assign a believability quotient of zero to the existence of *these* entities, and thus continue to be "agnostic about the existence of the [totally] unobservable aspects of the world described by science". If this unstated premise is granted, however, then Foss's critical conclusion is too modest. In the light of some of van Fraassen's other opinions, Foss's argument doesn't merely "reduce" the difference between realism and constructive empiricism—it *obliterates* it. In section 7.1, it was noted that van Fraassen explicitly repudiates the idea that theories can be associated with degrees of belief. He argues that if we admit that claims about unobservables have low but nonzero probabilities, then we can't resist the conclusion that there are actualizable circumstances under which these claims must become as near certain as makes no difference. Thus, the "Bayesian sort of constructive empiricist" is no constructive empiricist at all. When this issue first came up in section 7.1, I suggested that van Fraassen's objection to theoretical probabilities might be circumvented in some special cases. But the Bayesian sort of constructive empiricist that Foss envisions doesn't just want to ascribe probabilities to theories under special circumstances. She wants to do it across the board. The burden is on Bayesian antirealists to show why conditionalization on a small probability that electrons exist *couldn't* result in a high probability that they exist.

Van Fraassen's own response to the continuity argument is to reconceptualize Maxwell's series of graded cases from a continuum of *degrees of observability* to one of *proximity to a vague boundary*. Instead of molecules becoming increasingly less observable as they get smaller, it becomes increasingly uncertain whether they *are* observable. Observability itself is a completely dichotomous concept—everything either is or fails to be observable—but its boundaries are vague. Foss does not believe that this move helps van Fraassen to escape the foregoing objection. Indeed, his objection was directed to the vagueness interpretation in the first place. Presumably, the closer to the vague boundary an entity is supposed to be, the less likely it is that the entity *is* observable. Thus, the antirealist finds herself once again ascribing smaller and smaller degrees of credence to a series of claims that fall by degrees into the realm of the unobservable. The difference between this case and the previous one is that formerly the decreasing degrees of credence were assigned on the basis of decreasing degrees of observability; now they're assigned on the basis of decreasing probabilities that the event is (dichotomously) observable. But this interpretation leads to the same problem as before.

Or so Foss maintains. There is, however, an important difference between the two series of decreasing probabilities. Let's look at the graded series of entities that Maxwell presents us with—molecules in reverse size order—and ask how *the probability that the entity exists* falls off as we move from one end of the scale to the other. (It's assumed here that the entities all play equally essential roles in equally well-confirmed theories.) If what makes the probability fall off is the entity's decreasing degree of observability, then it's reasonable to suppose, as Foss does implicitly, that the decreasing function from entities to probabilities never quite hits zero probability: you can always take away a little bit more observability and lose a little bit more credence. But suppose instead that what makes the probability fall off is the entity's being more and more deeply immersed in the vague boundary between observability and unobservability. Then it's still reasonable to say that the probability gets smaller and smaller as it becomes increasingly uncertain whether the entity *is* observable— *but only until you get to entities on the other side of the boundary.* Once you get safely to the other side of a vague conceptual boundary, everything becomes clear again— the denizens of the other side are unambiguously *not* a part of the extension of the concept. Whereas the function from entities to probabilities in the first case was strictly decreasing and never hit zero, the function in *this* case contains a flat region on one side of the vague boundary where all the entities can be associated with zero probabilities (or with no probabilities, or with whatever it is that antirealists wish to ascribe to the existence of unambiguously unobservable entities). The existence of this region marks a clear difference between the antirealist and realist philosophies: there are some theoretical claims that antirealists will not allow themselves to believe, regardless of how well confirmed the theories they come from may be, but there's no such category of claims for realists. The fact that there is additionally a region of uncertainty doesn't erase this fundamental epistemic difference between the two camps.

The verdict: van Fraassen emerges unscathed from the continuity argument.

10.2 The Electron-Microscope-Eye Argument

Maxwell also claims that, quite aside from the gradual character of observability judgments, whether or not a particular entity has a given degree of observability changes with changes in the available instruments of science. For simplicity's sake, let's ignore the region of vagueness and suppose that it can clearly be established of every entity either that it's observable or that it isn't observable. Then the new problem is that much of what was unobservable in the past has now become observable because of the development of new scientific instruments, and that there's every reason to believe that some of the things that we're unable to observe at present will become observable by means of the new and improved observational technology of the future. If one equates observability with *current* detectability, observability becomes a context-dependent notion that won't sustain the antirealist thesis—at least not van Fraassen's version of the antirealist thesis. Van Fraassen doesn't just want to say that there's a class of entities that we can't believe in *now*—he wants to say that there's a class of entities that can *never* be believed in, no matter what happens in the worlds of science and technology. For this purpose, he has to formulate a concept of observability that's free of contextual dependencies. One alternative is to say that entities are unobservable if and only if they're undetectable by any nomically possible means of instrumentation. The problem with this formula is that there's no reason to believe that any entity postulated by science, if it exists, would fail to qualify as observable. The antirealist has no argument against the possibility that this concept of observability posits a distinction without a difference.

Van Fraassen tries to decontextualize the concept of observability in a different manner: he restricts the observable to what can be detected by the unaided senses. But does this move really effect a decontextualization? After all, Maxwell himself had brought up the possibility that human mutations might arise who have sensory capacities beyond our own. They might be able to observe ultraviolet radiation, or even X rays, with their unaided senses (1962, 11). Churchland (1985), mounting the same objection, asks us to consider the possibility of human mutants—or extraterrestrials—with electron-microscope eyes. Clearly, there's no more of a limit on the possibilities of genetic improvements in observational capacities than there is on technological improvements. Thus, the stipulation that observation be restricted to what can be accomplished by the unaided senses is no restriction at all.

Note that the same criticism can be leveled at Fodor's distinction: if you define "observability" as that which can be the output of an endogenously specified perceptual module, then there's no telling what may be deemed observable in the future, upon encounters with human mutants or extraterrestrials who have radically different perceptual modules. Churchland has used his electron-microscope-eye argument against Fodor (Churchland 1988) as well as against van Fraassen (Churchland 1985). Fodor's reply is similar to van Fraassen's (which I'll get to in a moment):

> Churchland apparently wants a naturalistic account of scientific objectivity to supply a guaranty that an arbitrary collection of intelligent organisms (for example, a collection consisting of some *Homo sapiens* and some Martians) would satisfy the empirical conditions for constituting a scientific community. *Of course* there can be no such guaranty. (Fodor 1988, 190)

A book could be written in explication of the notion of a "scientific community". For present purposes, the following rough-and-ready characterization will do: two beings are in the same scientific community if their opinions converge under ideal epistemic conditions. If, as seems likely to be the case, observability plays a special role in epistemology, then it may very well be necessary that two members of the same scientific community agree about what is observable. Fodor alerts us to the possibility that we and the Martians may fail to meet this requirement.

All this may be true as far as it goes. But it doesn't yet fully answer Maxwell's and Churchland's objection. It's possible that we and the Martians may be so differently endowed with senses that there can be no fruitful contact between our science and theirs. But it's also possible that we and the Martians are differently endowed with senses, but that the differences aren't so profound that there can be no fruitful scientific contact between us. The requirement that we agree on what is observable doesn't entail the requirement that we have the same sensory capacities. If A is able to observe a phenomenon that B can't observe, A and B may yet be part of the same scientific community. All that's required is that B be willing to *credit* A's observational reports about the events that B itself is unable to witness. After all, there are *blind* scientists who consider their *sighted* colleagues to be part of their scientific community. If we denied that one could ever regard as "observable" an event that we ourselves are unable to observe, then we would have to accuse such scientists of irrationality. I'm not prepared to spell out when it is or isn't appropriate to credit another's observational claims. But it seems sufficient for accreditation that there be a large amount of overlap between the two beings' sensory capacities, as there is between blind and sighted human scientists. In any case, whatever the crucial factor may be that allows blind and sighted scientists to be in the same scientific community, the same factor can surely be shared by sighted human scientists and mutant or extraterrestrial scientists. For instance, human mutants may develop whose sensory capacities are exactly the same as ours, except that they can see further into the ultraviolet spectrum than we can. It seems compelling that this case be treated the same as the blind-versus-sighted case: if it's rational for the blind to credit the sighted's visual reports, then it's equally rational for us nonmutants to credit the mutants' reports of ultraviolet perception.

But this is the first step onto a slippery slope. We've granted that any event is observable for *some* nomologically possible being, and that if the perceptual differences between us and other beings are sufficiently small, then it's rational to expand our scientific community to include them. Now consider a being M whose perceptual capacities are as different from ours as we like. There's going to be a *series* of nomologically possible beings that has the following properties: (1) its first member is us, (2) its last member is M, and (3) the perceptual differences between any two adjacent members in the series are so small that the rational thing for the nth being to do is to enlarge its scientific community so as to include the n-plus-first being. This means, of course, that for *any* supposedly "theoretical" entity X that exists, there are nomologically possible circumstances under which we have to admit that X is observable. And so both Fodor's and van Fraassen's concepts of observability posit a distinction without a difference.

If I'm right in my contention that Fodor doesn't really *need* a theory-observation distinction, then this conclusion makes no difference to his philosophical program. It makes an undeniable difference to van Fraassen's program, however. Yet, at least in 1980, van Fraassen seemed to accept the premise that our "epistemic community" (his term for Fodor's "scientific community") may change:

> It will be objected . . . that . . . what the anti-realist decides to believe about the world will depend in part on what he believes to be his, or rather the epistemic community's, accessible range of evidence. At present, we count the human race as the epistemic community to which we belong; but this race may mutate, or that community may be increased by adding other animals (terrestrial or extra-terrestrial) through relevant ideological or moral decisions ("to count them as persons"). Hence the anti-realist would, on my proposal, have to accept conditions of the form

> If the epistemic community changes in fashion Y, then my beliefs about the
> world will change in manner Z. (1980, 18)

But this admission leads van Fraassen into another version of the "Bayesian" objection discussed in section 10.1. Let X be an entity that figures in a well-confirmed scientific theory but that is currently unobservable by anyone in our epistemic community. The argument of the preceding paragraph shows that there's some nonzero probability that our epistemic community of the future will include members for whom X is observable. It follows that there's a nonzero probability that the hypothesis that X exists will rationally be granted a nonzero probability. But then the hypothesis that X exists must rationally be granted to have a nonzero probability *right now*. And then we're off to the races again: if it's granted that the hypothesis that X exists has even a ridiculously low nonzero probability, then van Fraassen himself tells us that we have to grant that there are actualizable circumstances under which we would have to grant that the hypothesis has a very *high* probability—as high as we require to say that we believe it. And therefore antirealism is false. Perhaps the argument is more perspicuous going the other way. Suppose I now think that it's impossible for us ever to find out whether X exists. Then I can't very well admit that our epistemic community may some day invite beings into it who can see Xs. So if I *do* admit this, I have to conclude that antirealism is false.

Clearly, what van Fraassen has to do to avoid the collapse of his antirealism is *not allow any flexibility in the composition of the epistemic community*. If you're in, you're in, and if you're out, you're going to stay out no matter what happens. That's the only way to assure there's going to be a class of claims that can *never* be believed, come what may. This goes counter to the suggestion of the passage quoted above from *The Scientific Image*. Van Fraassen seems to repeat the same suggestion in a reply to his critics a few years later:

> Significant encounters with dolphins, extraterrestrials, or the products of our own
> genetic engineering may lead us to widen the epistemic community. (Fraassen
> 1985, 256)

But on the next page, he seems to repudiate the idea that we may change our minds about the composition of the epistemic community. Speaking of the possibility that we might encounter extraterrestrials with electron-microscope eyes, he writes:

> In [this] case, we call the newly-found organisms "humanoids" without implying
> that they bear more than a physical resemblance to us. Then we examine them
> physically and physiologically and find that our science (which we accept as empirically adequate) entails that they are structurally like human beings with electron microscopes attached. Hence they are, according to our science, reliable indicators of whatever the usual combination of human with electron microscope
> reliably indicates. What we believe, given this consequence drawn from science
> and evidence, is determined by the opinion we have about our science's empirical adequacy—and the extension of "observable" is, *ex hypothesi*, unchanged.
> (256–257)

What van Fraassen has just given us is an entirely *different* argument for the contention that his antirealism entails that there can be no change in the composition of the epistemic community. My argument was that if you admit that the epistemic community can change, then you have to ascribe nonzero probabilities to what you think of as "theoretical" hypotheses *right now*. Van Fraassen's argument is that if the only part of your theory you believe in is the observational part, then you can never ascertain that beings outside of your current epistemic community actually *observe* anything. All you can tell is that they're reliable *indicators* of certain events, like the instruments that you use — and so you'll never encounter any reasons that rationally compel you to enlarge your epistemic community. So, there's really no question that antirealists have to be inflexible about who gets into the epistemic club.

Speaking technically, this conclusion spells *trouble* for antirealism. Suppose you're an antirealist who holds to the van Fraassian distinction between observables and unobservables. Then, it seems, you have to concede that the current boundaries of the epistemic community are unalterable for all time to come, regardless of whom or what you may encounter. This is big trouble, because it's a very big pill to swallow. After all, it's not as though van Fraassen or anybody else had offered us an epistemically relevant criterion for who should and who should not get included in the community in the first place. It's hard to imagine that there could be such a criterion that at once (a) is plausible, (b) allows blind and sighted scientists to be members of the same community, but (c) disallows the communality of the sighted scientists of the present and mutant scientists of the future who are just like them except that they report seeing a bit into the ultraviolet range. The fact that the boundaries include the blind and the sighted, but not the extrasighted, isn't *rationalized* in any way; it's presented to us as a fait accompli. In other words, the inflexible boundaries around the epistemic community are drawn *arbitrarily*. This is trouble indeed. But it isn't a devastation.

10.3 The Science-without-Experience Argument

Churchland wields another thought experiment against van Fraassen, which in some ways is the opposite of the electron-microscope-eye argument. The argument, inspired by Feyerabend's (1969) discussion of the possibility of "science without experience", runs as follows:

> Consider a man for whom absolutely *nothing* is observable. Fortunately, he has mounted on his skull a microcomputer fitted out with a variety of environmentally sensitive transducers. The computer is connected to his associative cortex . . . in such a way as to cause in him a string of singular beliefs about his local environment. These "intellectual intuitions" are not infallible, but let us suppose that they provide him with much the same information that our perceptual judgments provide us.
>
> For such a person, or for a society of such persons, the *observable* world is an empty set. There is no question, therefore, of their evaluating any theory by reference to its "empirical adequacy". . . . But such a society is still capable of science, I assert. They can invent theories, construct explanations of the facts-as-represented-in-past-spontaneous-beliefs, hazard predictions of the facts-as-represented-in-future-

spontaneous-beliefs, and so forth. In principle, there is no reason they could not learn as much as we have. (Churchland 1985, 42–43)

What does van Fraassen have to say about this argument? In response to Churchland's critique, he acknowledges that Churchland has given "several new twists" to Maxwell's mutant/extraterrestrial arguments that require further consideration. On van Fraassen's reckoning, Churchland's mutant arguments comprise the following:

First, there is the case of the man all of whose sensory modalities have been destroyed and who now receives surrogate sensory input electronically. . . . Next, he asks us to envisage an epistemic community consisting entirely of beings in this predicament. Thirdly, Churchland imagines that we encounter a race of humanoids whose left eyes have the same structure as a human eye plus an electron microscope. (van Fraassen 1985, 256)

He then proceeds to eliminate the middle argument from further consideration: "Upon reflection, it does not seem to me that the second example provides difficulties that really go beyond the third, so I will focus on that one" (256). But the second argument *does* provide difficulties that go beyond the third. Indeed, it's *only* the second argument that gives a new twist to Maxwell's discussion of a generation before. The third argument—the electron-microscope-eye argument—already occurs in Maxwell's paper (1962, 11). We've seen that the electron-microscope-eye argument already forces the antirealist to swallow the unpalatable conclusion that the boundaries of the epistemic community must be considered fixed for all eternity. Van Fraassen gives some indication of being willing to swallow this medicine. But the second argument—the science-without-experience argument—requires him to swallow even more. His response to the electron-microscope-eye argument commits him to excluding aliens with electron microscope eyes from his epistemic community, but it doesn't require him to deny that such aliens may be members of their *own* epistemic community, where a different but equally productive brand of science is carried out. It's just that we could never *know* that this was the case. In the case of beings with computers in their heads in lieu of sense organs, however, there's no question of their possibly having a science that's inaccessible to us. Van Fraassen's views commit him to saying that it's *impossible* for such creatures to have a science. For van Fraassen, science requires that some phenomena be observable, but his criterion for observability is detectability by the "unaided senses". Prosthetic devices don't count. Therefore, beings who rely entirely on prosthetic devices can have no science. It's not a matter, as in the previous case, of our not being able to *establish* that they have a science. They just *can't have one*. This new argument of Churchland's is not yet a devastation of van Fraassen's position. For one thing, it's conceivable that antirealists might be able to argue that such beings and their intellectual intuitions could not exist. But the argument has to be made. Churchland's objection makes new troubles for van Fraassen.

What about biting the bullet and conceding that creatures that have no sense organs can have no science? After all, what drives antirealism is the empiricist sentiment that the deliverances of our senses are the source of all knowledge. Why should an empiricist balk at accepting the conclusion that beings who have no sense organs can have no knowledge? The difficulty, of course, is that the beings described by Churchland are *computationally equivalent* to us in the strongest possible sense. To begin with, there's an isomorphism between the input-output relations that characterize us and those that characterize them. *Additionally*, the inputs into both types of systems are identical. It's light waves that go into our retinas, and it's light waves that go into their prosthetic devices. Thus, the equivalence of these systems is stronger than that which obtains between us and a computer simulation of us that parallels the input-output relations, but in which the inputs (light waves for us) are bit-map representations typed into a keyboard. Finally, the outputs of both types of systems are also identical—they're beliefs. Moreover, the beliefs of Churchland's creatures contain exactly as much of the truth as ours do. To those who still want to deny that such beings have a science, my temptation is to reply that what they call "science" just isn't very interesting. I'd rather talk about *schmience*, which is just like science except that its data are based on *schmobservation*, which is defined as observation or any other process that takes the same inputs as observations and is connected by isomorphic processes to the same outputs. For the remainder of this book, however— indeed, for the rest of my life—I will use "science" and "observation" as abbreviations for "schmience" and "schmobservation", respectively.

Does this move to schmience beg the question against antirealism? I don't think so. Antirealists still have the option of letting Churchland's "intellectual intuitions" count as observations. But then they have to give up van Fraassen's criterion of observability as detectability by the naked senses.

10.4 The Incoherence Argument

The next two arguments aim to devastate van Fraassen's distinction. The first of these is an attempt to precisify the obscure sense, already mentioned by Maxwell (1962, 9–10), that there's some sort of inconsistency between saying that one should believe no more of what a theory claims than that its observational consequences are true, and saying that it's our theories (which we don't believe) that tell us which consequences are observational. The possibility of such an inconsistency has been mentioned by a number of writers (Rosenberg 1983; Foss 1984; Giere 1985a; Musgrave 1985). The fullest treatment and most forthright accusation comes from Musgrave. According to Musgrave, "B is not observable by humans" isn't a statement about what is observable by humans. Thus, if a theory entails it, and I believe no more than that the theory is empirically adequate, then I have no grounds for believing that B is not observable by humans. But if I'm barred from establishing of any B that it's unobservable, then I can't, in practice, draw a distinction between the phenomena that are observable and those that aren't.

Van Fraassen has an adequate reply to this objection (1985, 256). He chooses to couch this reply in the model-theoretical terms that he favors for discussing the philosophy of science. It's worth paraphrasing it in a more familiar idiom, just to show that the mistake *isn't* due to the failure to think model-theoretically. Here's how it goes. Suppose B exists and is observable. Then if theory T entails that B is *not* observable, T will fail to be empirically adequate. So if we believe T to be empirically adequate, we have to believe either that B doesn't exist or that it's unobservable — equivalently, that *if* B exists, then it's unobservable. Musgrave is right when he claims that van Fraassen can't allow himself to believe that B is unobservable. But there's no reason why he shouldn't believe that B is unobservable *if it exists*. What antirealists refuse to believe is any statement that entails that theoretical entities exist. But the claim that theoretical entities are unobservable-if-they-exist doesn't violate this prescription.

The foregoing argument shows that it isn't logically inconsistent for antirealists to have hypothetical beliefs about entities whose existence is forever beyond our ken. But is this a plausible conclusion? Or should we regard an epistemology that generates this consequence to be in trouble? Hypothetical beliefs about nonexistent entities don't seem implausible to me. Imagine that all the evidence in a police investigation points to someone's having committed suicide. It might nevertheless not be irrational to believe that *if* the victim was murdered, the murderer must have been surpassingly clever. Of course, this observation doesn't completely settle the issue, for something might yet be made out of the differences between the two cases (e.g., antirealists think that there *can't* be any evidence for the existence of theoretical entities, whereas there simply happens not to *be* any evidence for the existence of a murderer). But I don't wish to press this point any further. We've found trouble enough for van Fraassen's distinction; one troublesome point more or less won't significantly alter its status. The important point here is that Musgrave's argument doesn't effect the devastation that it claims: it doesn't show van Fraassen's distinction to be incoherent.

10.5 Friedman's Argument

So far, van Fraassen's distinction has escaped with no more than some trouble. But the biggest hurdle is yet to come. Friedman (1982) claims that van Fraassen's position is incoherent because van Fraassen tells us (1) that we can believe in the observational consequences of our theories, and (2) that those consequences can only be expressed in theory-laden language. The problem is that if (2) is true, then every expression of our belief entails that some theoretical entities — namely, those posited by the theory we use to express our belief — exist. But this means that antirealism is false — for it's incoherent to hold that you can't believe in the logical consequences of hypotheses that you *can* believe. Friedman:

> "The observable objects" are themselves characterized from within the world picture of modern physics: as those complicated systems of elementary particles of the right size and "configuration" for reflecting light in the visible spectrum, for example. Hence, if I assert that observable objects exist, I have also asserted that certain complicated systems of elementary particles exist. But I have thereby asserted that (individual) elementary particles exist as well! I have not, in accordance with van Fraassen's "constructive empiricism," remained agnostic about the unobservable part of the world. (1982, 278)

The problem for van Fraassen is to find a formula that allows him to express what it is that his antirealism gives us license to believe, but that doesn't bring surplus theoretical entailments in its train. Here's what he had to say about this topic in 1980. In this passage, TN is Newton's theory of mechanics and gravitation, which includes the hypothesis of absolute space, and TN(v) is the theory TN plus the postulate that the center of gravity of the solar system has constant absolute velocity v. Van Fraassen writes: "By Newton's own account, he claims empirical adequacy for TN(o); and also that, if TN(o) is empirically adequate, then so are all the theories TN(v)" (46). The question is, "what exactly is the 'empirical import' of TN(o)?":

> Let us focus on a fictitious and anachronistic philosopher, Leibniz*, whose only quarrel with Newton's theory is that he does not believe in the existence of Absolute Space. As a corollary, of course, he can attach no "physical significance" to statements about absolute motion. Leibniz* believes, like Newton, that TN(o) is empirically adequate; but not that it is true. . . . What does Leibniz* believe, then?
>
> Leibniz* believes that TN(o) is empirically adequate, and hence equivalently, that all the theories TN(v) are empirically adequate. Yet we cannot identify the theory which Leibniz* holds about the world—call it TNE—with the common part of all the theories TN(v). For each of the theories TN(v) has such consequences as that the earth has *some* absolute velocity, and that Absolute Space exists. . . .
>
> The theory which Leibniz* holds of the world, TNE, can nevertheless be stated, and I have already done so. Its single axiom can be the assertion that TN(o) is empirically adequate. . . . Since TN(o) can be stated in English, this completes the job. (46–47)

The claim is that, relative to any accepted theory T, the content of what antirealists believe is precisely and fully specified by the formula "T is empirically adequate". Nothing more is needed. The question is whether this formulation avoids Friedman's problem. I'd like to discuss this question twice—once using van Fraassen's informal and preliminary definition of empirical adequacy, and a second time using his official definition. My conclusion is that it makes no difference here (as in most places) which of the two definitions is used, whereupon I revert to the more familiar preliminary notion for the rest of the discussion.

The informal notion is that a theory is empirically adequate if and only if all of its "observational consequences" are true (van Fraassen 1980, 12). Moreover, the existence of an entity X is an observational consequence of theory T if (1) the existence of X is a consequence of T, and (2) X's being observable is also a consequence of T (or of T in conjunction with other theories that we accept). This formulation leads ineluctably to Friedman's problem. It's a consequence of our current theoretical views that certain structures comprising more than 10^{23} atoms of carbon are observable and that they exist. That is to say, the existence of these structures is an observational consequence of current theory. Thus, someone who believes current theory to be empirically adequate is committed to believing that certain structures of carbon atoms exist. But this belief in turn entails that individual carbon atoms exist. Therefore, antirealism is false. If we're to avoid this three-step devastation, we need a narrower notion of "observable consequences"—a notion of *pure* observable consequences—according to which the existence of macroscopic lumps of carbon is a pure observational consequence, but the existence of individual carbon atoms isn't. Since van Fraassen himself *tells* us that there's no theory-neutral language, how will we ever be able to pull off such a trick?

What about admitting that we can't *say* what the pure observational consequences of our theories are, but that we *know* what they are nonetheless? For what it's worth, my intuition is that it is possible to make sense of the idea of ineffable knowledge. But, be that as it may, the enterprise of *science* is certainly not in the business of promulgating ineffable knowledge. Effing is what science is all *about*. If we allow that scientists may, in their professional capacities, believe unspecified portions of theories, then it's not clear that the theories themselves play any essential role. After all, whether or not we admit that there are inexpressible beliefs, every possible state of belief is equivalent to belief in a portion of the universal theory U which claims that every hypothesis is true. To be sure, U can't itself be true, since it makes contradictory claims. But there's nothing that a scientist might believe that isn't a *part* of U. So if it's allowed that the result of scientific research can be a state of belief in an unspecified part of a theory, then we have to admit that the *final* result of scientific research—the state wherein we're prepared to say that the work of science is *finished*—can be a state of belief in an unspecified portion of U. But this is to say that the formulation of theories is optional in the first place. Compare the following rule for finding a number: start with the number 6, double it, add 9, divide by 3, then subtract any number you like. Obviously, you could produce the same effect by skipping the "rule" altogether and just telling people to pick a number. So it is with the idea that we can believe in unspecified portions of theories. If van Fraassen were to avail himself of the idea that the purely observational consequences of a theory are ineffable, he would lose all semblance of an account of the role of theories in science.

Now let's look at van Fraassen's *real* definition of empirical adequacy. What van Fraassen would surely say of the foregoing disquisition is that the need to describe the empirical content of our theories as a set of sentences is a hangover from the "syntactic" view of a theory as a set of axioms closed under the operation of logical entailment. The equation of empirical adequacy with the truth of a theory's "observational consequences" is a temporary crutch for the syntactically afflicted, to be discarded upon the acquisition of the *semantic* story about empirical adequacy. This story, in brief, is that a theory is to be equated with a set of models, and that to say that a theory T is empirically adequate is to say that the world of observable phenomena is isomorphic to a submodel of T (van Fraassen 1980, 64). On this account, the belief that our current theories are empirically adequate does *not* entail the truth of theoretical assertions that make reference to structures comprising 10^{23} carbon atoms. Strictly speaking, these *and all other* sentences formulated in the language of the theory have no truth-values at all. What may be true or false is the belief that the circumstance in the world coordinated with the theoretical statement about carbon atoms obtains. But, patently, to say this is simply to shift the locus of Friedman's problem to another place. The question now is: *Which* event in the world is the one that's coordinated with the theoretical claim about 10^{23} carbon atoms? To *use* the atomic theory, we have to specify the rules of coordination that tell us which items in the world correspond to which items in the theory. Giere (1985a), who agrees with van Fraassen about the superiority of the semantic approach to theories, calls such coordinating rules *theoretical hypotheses*. How are these theoretical hypotheses to be expressed? If we had a theory-neutral language in which to describe the observable phenomena, we could coordinate the entities posited by the theory with phenomena in the world by using theoretical language for the former and theory-neutral language for the latter. But van Fraassen tells us that there is no theory-neutral language. In that case, the only way to specify which observable items in the world correspond to various items in the theory is by using *some* theoretical description of the observable items in the world — and that commits us to the existence of the theoretical entities referred to in that description. This is the first horn of the same dilemma that was encountered with the syntactic view: it seems impossible to assert what we want to assert without making theoretical commitments. The second horn is that it's also unacceptable to repudiate the need to say it. To claim that the world of observable phenomena is isomorphic to a submodel of our theory without being able to give the coordinating rules between elements of the world and elements of the theory is obviously to leave the theory uselessly suspended above the world.

Friedman's argument devastates van Fraassen's distinction. If we define observability as van Fraassen does, it follows that antirealism is false. Antirealists would be better off with *Fodor's* concept of the observable as the outputs of endogenously specified perceptual modules. To be sure, this concept was found to be troublesome for antirealists, since no one has shown how to make it relevant to their epistemic thesis. But antirealists were going to have trouble with van Fraassen's notion anyway, because of the electron-microscope-eye and science-without-experience arguments. An even more serious shortcoming of Fodor's concept is that it isn't viable in the Kuhnian model of perception, according to which there *are* no endogenously specified perceptual modules. Thus, if antirealists take the Fodorean route, the coherence of their epistemic thesis will hang on unsettled empirical questions in the psychology of perception.

Perhaps an entirely new way of drawing the theory-observation distinction would avoid both Friedman's problem and the dependence on chancy empirical results. But until such a distinction is forthcoming, it would be no more than an act of faith on the part of antirealists to suppose that the coherence of their position will be vindicated by the invention of a new concept of observability. Moreover, since the distinction hasn't even been made yet, it would be highly irrational for antirealists to suppose that, when it is finally made, it *will* lead to the consequence that antirealism is the best epistemic policy.

The Theory-Observation Distinction III

The Third Distinction

Here's a way of drawing the theory-observation distinction that seems to avoid both the devastation of van Fraassen's distinction and the chancy dependence on empirical results of Fodor's distinction. Let's suppose that theories are formulated in such a way that their singular consequences are about the occurrence or non-occurrence of events. Van Fraassen, for his part, is willing to admit that statements about objects are intertranslatable with statements about events: "There is a molecule in this place" and "The event of there-being-a-molecule occurs in this place" are merely notational variants (van Fraassen 1980, 58). Let OX be the proposition that an event of type X occurs. If X is the decay of a particle of type A, then OX is the proposition that an A-particle decays. Also let E(T, "X") be the event (type) that theory T refers to as "X". If T is true, then E(T, "X") = X. For instance, if current particle physics is true, then it's also true that the event that current particle physics refers to as "the decay of an A-particle" *is* the decay of an A-particle. But this identity may fail for false Ts. Consider, for example, the false theory that identifies certain types of shimmering reflections with the presence of ghosts. Then, a particular type of reflection may very well be the type of event that this theory refers to as "the presence of a ghost", but the reflection wouldn't *be* a ghost.

Now define the *atomic observation sentences* as all sentences of the form OE(T, "X"), where T is any theory and X is any event. If T is current particle physics and X is the decay of a particle of type A, then OE(T, "X") is the following observation sentence: "An event takes place that particle physics refers to as 'the decay of an A-particle'". The complete class of observation sentences is generated by closing the set of atomic observation sentences under truth-functional operations and quantification.

Note, to begin with, that this distinction partakes of some of the properties of Fodor's distinction and some of the properties of van Fraassen's. Like Fodor's and his positivist predecessors', it's a distinction between types of *sentences* rather than between entities or events. Like van Fraassen's, however, it doesn't try to characterize the difference between observation statements and nonobservation statements in terms of two discrete subvocabularies. Rather, observation sentences are those that meet a certain conceptual criterion. As in van Fraassen's distinction, the difference between the observable and the unobservable can't be expressed in theory-neutral language. But unlike van Fraassen's distinction, the theory-laden language used to describe the observable occurs in an oblique context.

The most important characteristics of the third distinction are that it manages to avoid both the devastation of van Fraassen's distinction by Friedman's argument and the dependence on strong empirical results of Fodor's distinction. First, Friedman's problem: if OX is a consequence of T (together with the appropriate initial conditions and auxiliaries), then OE(T, "X") is also a consequence of T; therefore, since OE(T, "X") is an observation sentence by definition, OE(T, "X") is an *observable consequence* of T. Nevertheless, *OE(T, "X") does not, all by itself, imply* OX. If it's a consequence of theory T that a particular setup will reveal an electron track, then it's a consequence of T that the setup will produce an event that T refers to as an "electron track". But the fact that a particular setup produces an event that T refers to as an "electron track" doesn't entail that the event *is* an electron track. Thus, Friedman's problem is averted.

Second, unlike Fodor's distinction, this one is viable in all three models of perception—the Fodorean, the Kuhnian, and the mixed. Antirealists who adopt the new distinction won't have to hold their breath and hope that the modular theory turns out to be true. In fact, the third distinction doesn't make any heavy-duty structural assumptions about our cognitive-perceptual machinery. Even Churchland's race of beings with computers in their heads in lieu of sense organs could indicate the event that causes them to have a particular "intellectual intuition" by the formula OE(T, "X"). This means, incidentally, that the third distinction avoids the trouble that the science-without-experience argument makes for van Fraassen's distinction. There's no reason why antirealists who employ the third distinction shouldn't admit that there can be a science without sensory experience.

In fact, the third distinction manages to avoid all the old troubles that afflict its predecessors. For example, antirealists who want to avail themselves of Fodor's distinction face the problem of its *arbitrariness*: there seems to be no connection between their epistemic concerns and the question of whether a particular claim directly reflects the output of the perceptual module. In the case of the third distinction, however, there's at least a prima facie reason to suppose that OE(T, "X") is on a firmer epistemic footing than OX. Assume it's true that every sentence is ineliminably theory-laden. This is, after all, the assumption that's making all the trouble: if some sentences were theory-free, then antirealists could avail themselves of the old positivist notion of a pure observation language. Now, consider the hypothesis OX. By the assumption about theory-ladenness, to assert that this hypothesis is true is to take the standpoint of some theory T. More precisely, it's to presuppose that T, the theory from which standpoint we judge that OX is true, is correct at least to this extent— that OX is true. Now I claim that the hypothesis OE(T, "X") is *logically weaker* than OX. It's not that OX is logically stronger than *any* claim of the form OE(T, "X"). For instance, it's possible that OX is true, but that some theory T so misdescribes phenomena that the event that *it* (erroneously) refers to as X doesn't occur—that is, that OE(T, "X") is false. So OX doesn't entail OE(T, "X") for *every* T. But OX *does* entail OE(T, "X") *for the T that's being presupposed when one makes the claim* OX. Having taken the standpoint of T, we may be impelled by the evidence to claim that OX.

But if we content ourselves instead with asserting that OE(T, "X"), we make a weaker claim: we don't claim that T is right about the occurrence of X—we claim only that something occurs that T refers to as "X". There are more ways of going wrong with the claim that a subatomic particle A decays than with the claim that an event occurs that our current theories refer to as "the decay of an A-particle". This is the kind of difference that might very well have a bearing on our epistemic assessments. It remains to be seen whether this difference can be parlayed into a case for believing one category of claims but refusing to believe the other. But at least it's sufficient to allow the antirealists to graduate to the second round of the debate.

Next, consider the trouble made for van Fraassen's distinction by the electron-microscope-eye argument. The problem here is that the contemplation of electron-microscope-eye beings leads us to the conclusion that *any* event might be observable. The only way that van Fraassen can avoid the conclusion that his distinction is a distinction without a difference is by permanently closing the borders of the epistemic community. If we use the third distinction, however, we can afford to admit that every event is observable, in the sense that the occurrence of any event can be described by an observation sentence. While this admission would devastate van Fraassen's distinction, it has no effect on the third distinction, because the latter isn't a distinction between *events*—it's a distinction between *sentences*. It may be true that the occurrence of any event can be described by an observation sentence. But that doesn't affect the distinction between observation sentences and nonobservation sentences. "A virus of type A disintegrates" isn't an observation sentence, even for beings who have electron-microscope eyes with which to see the putative virus disintegration "directly". "What current virological theory refers to as 'the disintegration of a virus of type A' is happening" *is* an observation sentence, even if we *don't* have electron-microscope eyes. In fact, it's an observation sentence even if we don't have electron microscopes. We can tell that it's an observation sentence on the basis of its conceptual properties alone.

So, the third distinction is free of all the old troubles and devastations that afflict its predecessors. This does not, of course, ensure that it isn't heir to new afflictions. In fact, somebody has tried to make a new problem for it already. The problem arises when we ask the following question: What determines the extension of a term like E(T, "X")? If we believe that theory T is true, the answer is easy: if somebody wants to know what E(T, "X") is, we can tell her that it's X. What theory T refers to as an "electron track" is neither more nor less than the track of an electron. But obviously, this means of conveying the extension of E(T, "X") isn't available when T is a theory that we don't believe. In these cases, it seems clear that the only available option is to suppose that the extensions of such terms are fixed *demonstratively*. Antirealists can't say that what theory T refers to as an "electron track" is an electron track. If they're logical positivists, they may try to express what theory T refers to as an "electron track" in theory-neutral observation language. But if, like van Fraassen, they believe that all language is theory-laden, then the only thing they can say is that what theory T refers to as an "electron track" is *this*. Traveling along a somewhat different argumentative path, Stephen Leeds (1994) has also come to the conclusion that the only way we might be able to fix the extensions of the terms denoting observable events in theories that we don't believe is via demonstratives. However, he claims that this procedure is doomed to failure. Let G be an observable entity or event postulated by theory T:

> The difficulty is that . . . the extension of G will often go well beyond the situations in which an ordinary mortal is able to say "Here is a G": and it is not easy to see a determinate way to project from one to the other. . . . Take an example: suppose that the speakers of T have a word H which they apply pretty regularly to observable situations in which hydrogen is present. Pretend that at temperatures at which humans can exist, hydrogen is never ionized, and suppose that in the interior of the sun hydrogen *is* ionized. Then what makes it the case that the hydrogen of the sun is—or isn't—in the extension of H? (195–196)

My response is that the difficulty to which Leeds alludes is far too general to bring to bear against the coherence of antirealism. In fact, the difficulty is of the same order as Goodman's new riddle of induction. The fact that some members of the denotation of H can't be pointed to because of our physical limitations (we can't point to things in the interior of the sun) is incidental to the problem. What makes a problem is that *we can't exhaustively point to all the instances of our concepts*. It doesn't matter whether the omitted instances are omitted because of physical disability, or merely because of lack of time. In either case, there are indefinitely many mutually incompatible ways of projecting the concept onto the omitted instances. To paraphrase Leeds: having pointed to a million instances of green things, what makes it the case that the extension of "green" comprises green things rather than grue things? The question could also be put in Kripkensteinian terms: given the first *n* terms of a series, how is it that two people ever continue it the same way? There are only two classes of answers to this question—the Wittgensteinian and the Chomskian. The Wittgensteinian answer is that they just do—"explanation has to stop somewhere". The Chomskian answer is that the same continuation is due to shared endogenous constraints. For what they're worth, both of these solutions are available for dealing with Leeds's scenario. I'm not claiming that either of them is satisfactory. But their possible inadequacies can't be used as an argument against the coherence of antirealism, because just about *everybody* needs to assume that we can fix the extension of a term by pointing to a finite number of its members.

Leeds thinks otherwise. He claims that the problem is acute only for antirealists:

> Of course this sort of projection problem isn't uniquely a problem for the constructive empiricist. There will also be cases in which even a Realist . . . will be unable uniquely to project the extension of a predicate of T. The difference is that, for the Realist, this project is not one at which he needs to succeed. He is not committed to producing a unique observational content for arbitrary theories T. Only for the theories that he himself believes will he expect to say that each term has a determinate extension: he will, for example, say that the extension of "hydrogen" is the set of everything which is hydrogen—and of course the Realist will be able to call on the resources of his own theory to explain what it is to be hydrogen. (196)

The argument is that realists don't need to rely on ostensive definition to specify the extensions of the terms of the theories they believe in, because these extensions can be given verbally: "hydrogen" means hydrogen. Antirealists can't avail themselves of this form of specification because they don't believe that there *is* such a thing as hydrogen. So realists don't need ostensive definition for *that* purpose. But surely realists have more epistemic business to conduct than merely to give a verbal description of the semantics of the theories they believe in. For one thing, they might want to know how to confirm or disconfirm theories that posit alternative ontologies. Even more indispensably, they will want to know how their own language hooks up to the world. It's true that if you're a realist, you can answer the question "What does the term 'hydrogen' refer to?" by saying that "hydrogen" refers to hydrogen. But a complete specification of the semantics of T in this style is like the proverbial unabridged dictionary in the hands of a Martian. There's a sufficiently broad, but still natural, sense of the word "point" such that it's correct to say that you can't learn to talk about the world without having some things pointed out to you at some stage. While it may be true that antirealists have a *particular* need for ostensive definition to ground their epistemic thesis, and that realists don't have *that* need, it's not true that realists can conduct their affairs in a normal manner without relying on ostensive definition at any point. Even the Kuhnian relativists need ostensive definition to fix the extensions of the primitive terms of their paradigms—the fact that one Kuhnian's primitive terms are incommensurable with another's is neither here nor there. Just about everybody needs ostensive definition. As a consequence, just about nobody is in a position to use the problems and puzzles relating to ostensive definition as arguments against their philosophical opponents.

Regardless of whether we can explain how it's done, it seems obvious that we do, in fact, manage to discriminate situations that are deemed to be relevantly different by theories that are false, or that we don't believe in. There are several ways to see this. For example, there's the fact that proponents of now-defunct theories such as the phlogiston theory or the ether theory were able to make the perceptual discriminations necessary for confirming or disconfirming hypotheses relating to their theories. This shows that the capacity to make theoretically relevant discriminations doesn't depend on the truth of the theory. Moreover, when the contemporaries of Lavoisier *gave up* their belief in the phlogiston theory, it can hardly be supposed that they suddenly lost their capacity to make the same discriminations as before. Nor, supposing that current particle physics were to be utterly discredited, does it seem at all plausible to expect that physicists of the transitional generation will lose their capacity to discriminate what they used to call "electron tracks". This establishes that the capacity to make a theoretically relevant discrimination doesn't depend on our *believing* that the theory is true, either. So there really is no problem about fixing the extension of E(T, "X")—or at least no problem that carries any dialectical weight.

In sum, the third distinction suffers from no *published* troubles or sources of devastation. But, of course, nobody has yet had the opportunity to look very hard for troubles and devastations. No doubt there are conceptual land mines lying all around this territory. Here are a few things to worry about. The list is decidedly not exhaustive.

To begin with, there's the fact that the third distinction does presuppose that *some* language is theory-neutral—namely, the concepts of an event and of its happening, the apparatus needed for the wielding of demonstratives, the machinery of semantic ascent, and locutions such as "refers". To my mind, these items are on a par with sentential connectives and quantifiers, which, I think, even the Azande regularly employ. My feeling is that Kuhnians ought to be able to live with this amount of theory-neutrality.

Second, there's the possibility that the third distinction might be entangled even more deeply and more problematically with counterfactuals than is usually the case with scientific theories. In the case of run-of-the-mill counterfactuals, such as "If this match were struck, then it would light", we inquire into the goings-on in nearby possible worlds where the laws of nature are the same as they are here. These run-of-the-mill cases are hard enough to understand. But doesn't the third distinction require us to journey to even more distant possible worlds, where even the laws of nature have been changed? For doesn't OE(T, "X") mean that if T were true, then the event designated by X would occur? Maybe. Or maybe OE(T, "X") can be interpreted in terms of run-of-the-mill counterfactuals about psychology — something on the order of: if a perceptual system were assembled that embodied the principles of T, its output would be that OX.

Next, there's the problem of productivity. We don't want to have to learn the meaning of every new observation statement ostensively. After a while, we have to be able to *figure out* the meanings of observation statements about events we've never encountered before. How do we do that? What are the laws of composition for expressions of the form OE(T, "X")?

There's another type of problem relating to the third distinction. I'm sure it's been preying on the attentive reader's mind for some time. Grant that the distinction is coherent and that it's relevant to our epistemic concerns. Even so, it might not provide a passport for antirealists to migrate to the second round of argumentation, for it might not be a *theory-observation* distinction. We've already seen that the sesquicentenarist distinction between the first 150 mutually consistent hypotheses and the rest wouldn't help the antirealists' cause, even if it could be shown that sesquicentenarism is true. The reason is that this distinction doesn't make any connection to the antirealists' empiricist motivations for their skepticism about theoretical entities — equivalently, it isn't a theory-observation distinction. Does the third distinction fare any better in this regard? I think it does, but I'm not yet able to present a compelling case for this opinion. For the present, I'll simply add this potential problem to the list of issues in need of further attention. However, here's a preliminary (i.e., not compelling) case for considering the third distinction to be a theory-observation distinction.

First, it gives the right result in placing the paradigmatic cases of "theoretical statements" — for example, "electrons are streaming through the cathode ray tube" — in the nonobservational category. This criterion would already eliminate the sesquicentenarist distinction from the ranks of potential theory-observation distinctions. One might object that the third distinction just as clearly gives the *wrong* result for paradigmatic *observation* statements like "the cat is on the mat", for these are also to be relegated to the realm of the nonobservable. I deny that the two cases are on a par. If you think, as van Fraassen does, that the putative observation languages of Fodor and the logical positivists are "thoroughly theory-infested," then "the cat is on the mat" has been placed where it belongs.

Statements like OE(T, "X"), on the other hand, are arguably uninfected by theoretical commitments. They're what you get when you try to take the theoretical commitments *out* of your claims. Standing in front of a cathode ray tube, an observer might say "there's a stream of electrons." She might even be right. But her being right requires that a theory of physics be at least partially right. On the other hand, if she says "there's an event that contemporary physics refers to as 'a stream of electrons'", the correctness of her claim depends only on (1) the nature of her perceptual experience and (2) her understanding of the conceptual machinery of modern physics. Aren't we getting very close here to what empiricism is all about? To a first approximation, the empiricist root notion of an "observational" claim is that of a hypothesis whose rational acceptance can be based entirely on perceptual experience (together with the requisite conceptual knowledge). I would recommend to empiricists that they liberalize this notion in a way that accommodates the computer-in-the-head beings who do science—or schmience—without enjoying any perceptual experiences. Instead of requiring that the observational be based exclusively on perceptions, we could say that it's based on what is *given* to us. For beings like ourselves, the given is delivered via perceptual experiences; for computer-in-the-head beings, it comes in the form of "intellectual intuitions". In neither case, however, is the given properly reported by *endorsing* the perception or the intellectual intuition. When we see that the cat is on the mat, the given isn't that the cat is on the mat—it's that we *see* that the cat is on the mat—in other words, it's that an event takes place that the theory embodied in our perceptual system refers to as "the cat is on the mat". Similarly, when computer-in-the-head beings have the intellectual intuition that the cat is on the mat, the given is that an event takes place that the theory embodied in their implanted computer's program refers to as "the cat is on the mat".

Granting that the third distinction's concept of the observational doesn't exclude too much when it excludes "the cat is on the mat", what about the charge that it excludes too little? If current physical theories are correct, then some statements that refer to the collision of elementary particles are going to turn out to be observation statements. Does this do violence to the empiricist's intuitions? I think not. Recall that one problem with van Fraassen's concept of the observational was that it had to be admitted that mutant or extraterrestrial scientists may be able to perceive events such as the collision of elementary particles. It requires fancy philosophical footwork to avoid the conclusion that any event is potentially detectable by the unaided senses. I suggest that we forego the footwork and admit that this is so. The third distinction provides us with a nonvacuous conception of the observational that can coexist with this admission.

Let's set the apologies aside and simply get clear on what follows from each of the several ways in which the outstanding problems with the third distinction might be resolved. There are three hypotheses to consider: (1) the third distinction is coherent; (2) the third distinction is a theory-observation distinction; (3) the third distinction divides scientific hypotheses into a class that may rationally be believed under some actualizable circumstances and a class that can never rationally be believed (this is the hypothesis that gets discussed in the second round of debate). The investigation into each of these hypotheses may have any of three outcomes: accept the hypothesis, reject it, or come to no definite conclusion. Moreover, the acceptance of either (2) or (3) entails the acceptance of (1). Here is what all possible outcomes of these investigations would entail for realism, antirealism, and relativism. To begin with, if *any* of the three hypotheses is rejected, then van Fraassen's brand of epistemic antirealism is as close to finished as it can ever be—for antirealism needs a distinction that has all three of these properties, and the third distinction is the only candidate left on the table. Of course, it would still be possible for some as yet undreamed-of interpretation of observationality to rescue antirealism from oblivion. But the same can always be said of any doctrine that allows for some interpretative latitude. What one does in a situation like this is partly a matter of philosophical taste. There are many who would feel that there's no need to pay any further attention to antirealist arguments unless and until the undreamed-of vindicatory concept materializes. I don't think that this attitude can be faulted. But neither would it be a mistake to investigate the second-round case for antirealism, wherein arguments about the epistemic differences between the theoretical and the observational are exchanged, under the *supposition* that we have a coherent theory-observation distinction. Of course, it may turn out that the ultimate disposition of some of the second-round arguments will depend on how, precisely, the distinction is drawn. There are bound to be some limits on the work that can be done with a placeholder in lieu of a real concept. But the first seven chapters of this book bear testimony to how much of the second-round argumentation is unaffected by this limitation.

If all three hypotheses about the third distinction were to be *accepted*, then of course antirealism would win. This is the only possible outcome of the investigation of hypotheses (1), (2), and (3) that would eliminate realism. Relativism loses if the third hypothesis is accepted—for this hypothesis tells us that we can have absolute warrant for believing in some claims, whereas relativists maintain that we are never absolutely warranted in believing any claims. Thus, a combination of rejecting hypothesis (2) and accepting (3) would produce a clear victory for realism: the rejection of (2) would eliminate antirealism, and the acceptance of (3) would rule out relativism.[1] This state of affairs is akin to the one that would obtain if sesquicentenarism were shown to be true. Antirealism is ruled out, and realism and relativism are left in the running, if (3) is not accepted and at least one of the three hypotheses is rejected— for example, if it's established that the third distinction is coherent and that it's a theory-observation distinction, but it turns out that it has no epistemic import. Finally, all three contenders remain in the running if (3) isn't accepted and nothing is rejected— for example, if it's established that the third distinction is coherent and that it's a theory-observation distinction, but no decision can be reached as to whether it has epistemic import.

The third distinction has all the vices and all the virtues of the inchoate. Its main vice is that it might very well suffer from glaring deficiencies that have yet to be unearthed. Its main virtue is that these deficiencies haven't yet been unearthed. My view is that the status of all three key hypotheses relating to this distinction remains unsettled, so all three contenders—realism, antirealism, and relativism—are still in the running. In light of their close brush with disaster, that would have to be considered a happy conclusion for antirealists.

Realism and Epistemology

12.1 The History of Antirealism in the Twentieth Century

The history of antirealism in the twentieth century exhibits a pattern of progressive attenuation, wherein each version of the doctrine lays claim to strictly less than its predecessor. Van Fraassen's 1980 brand of antirealism—"constructive empiricism"—already represents a retreat from the bold claims of the positivists. Constructive empiricists concede that statements about theoretical entities can't be reduced to statements about actual or possible observations and that such statements are nevertheless true or false. In the terminology introduced in chapter 1, they're both antireductionists and semantic realists. What marks constructive empiricism as a brand of antirealism is the epistemic claim that we're never warranted in believing more of a theory than that it's empirically adequate. Strictly speaking, the view that theories have truth-values isn't logically weaker than the positivists' assertion that theories *don't* have truth-values. But it's clear that van Fraassen doesn't *need* semantic realism to establish what he wants to say about science. If it were shown that theories are *not* true or false, he would be able to greet the news with equanimity. Indeed, it would lighten his argumentative burden. Van Fraassen merely concedes the point about truth-valuability because he thinks he can say what he wants about science despite this concession. Thus, it's more appropriate to regard constructive empiricism as claiming that theories *may or may not* have truth-values, for this is all that van Fraassen needs. And this doctrine certainly is weaker than the positivists'.

Between 1980 and the 1989 publication of *Laws and Symmetry*, van Fraassen's antirealist claims undergo an additional attenuation. In 1980, constructive empiricism is presented as a conclusion that follows from arguments that ought to persuade any rational person to abandon realism. In a 1985 reply to his critics, van Fraassen equivocates between the relatively strong claims of 1980 and the permissive turn in his epistemology that is to come. (The pronouncements of this intermediate van Fraassen were examined in section 7.2.) By 1989, van Fraassen explicitly concedes that it isn't irrational to be a realist. His claim is only that it also isn't irrational to be an antirealist. Before treating these later developments, let's briefly summarize the fate of the classical constructive empiricism of 1980.

In *The Scientific Image*, van Fraassen seems to be saying that the epistemic superiority of constructive empiricism over realism can be established on grounds that rationally compel the realist's assent. He gives an argument from scientific practice that purports to show that constructive empiricism provides a better account of actual scientific practice than realism does. There are also fragments of the argument from underdetermination scattered throughout the book. These arguments were discussed in previous chapters and were found to be inconclusive at best. The failure of the 1980 case for antirealism is ultimately due to its reliance on the Vulnerability Criterion of Belief, the principle that we should reject the stronger of any two hypotheses that are equally vulnerable to disconfirmation by experience. The problem with this principle is that antirealists don't have the philosophical resources needed to compel its acceptance by the realists. In fact, the most common reason for becoming a scientific realist in the first place is that one thinks that some nonempirical property of theories such as their explanatoriness or simplicity can be a warrant for belief. As far as the arguments of 1980 go, van Fraassen begs the most important question of all against the realists.

As long as we're reviewing established results, it should also be recalled that realists have begged the question just as often the other way by simply assuming that some nonempirical virtue like explanatoriness or simplicity *is* epistemically significant. This happens, inter alia, in the miracle argument. The premise of the miracle argument is that the truth of our theories provides us with the only explanation for their empirical success. This is supposed to be a reason for adopting the realist belief that our successful theories *are* true. However, the movement from the premise to the conclusion requires us to assume that explanatoriness is an epistemic virtue — and this is just what an antirealist like van Fraassen denies. Therefore, the miracle argument is no more telling against antirealists than the underdetermination argument is against the realists.

In sum, the critical *counter*arguments on both sides of the debate have been considerably more successful than the positive arguments. They've swept all the positive arguments away, leaving an empty field. It's this impasse that has prompted developments like van Fraassen's later and more permissive stance, the natural ontological attitude (NOA), and various other relativisms and postmodernisms. It's time to face up to the possibility that the debate is irreconcilable in principle.

12.2 The Irreconcilability of the Debate

There are passages in chapter 7 of *Laws and Symmetry* that sound very much as though van Fraassen were conceding that scientific realism is no worse a choice than antirealism. He explicitly grants that it isn't irrational to be a realist. The only question, then, is whether he claims any objective inferiority for realism short of irrationality, or whether the choice is conceded to be purely a matter of personal preference. In an attempt to elucidate his stance, van Fraassen likens the philosophical choice between realism and antirealism to making a political commitment:

Suppose . . . [we] are . . . committed that racism or poverty shall end, that the homeless shall be fed, that war shall be no more. Any such commitment *may* be subject to a telling critique. In taking the stance, we avow that we have faced all reasons to the contrary. But we do not pretend that, if we had historically, intellec- tually, or emotionally evolved in some different way, we might not have come to a contrary commitment. We accuse this alternative history . . . of being mistaken or wrong (just like our opponents) but not of being irrational. (1989, 179)

There are two ways to understand the claim that our opponents are "mistaken or wrong" but not irrational. This characterization might indicate that we disagree with our opponents but that we're unable to present them with reasons that should com- pel them to change their ways. Alternatively, when we characterize our opponents as "mistaken", we may want to suggest that they've committed a misstep that they're in principle capable of recognizing and rectifying. For example, we might think they've neglected to take a crucial piece of information into account. But the dis- tinction between this kind of mistake and being irrational doesn't seem very impor- tant. In the debate between scientific realists and constructive empiricists, what does it matter whether the wrong side believes what it does on the basis of an invalid philosophical argument, or because it's wrong about the history of science? Either way, it loses.

Which kind of wrongness does van Fraassen attribute to his realist opponents? I think that the passage can be read either way. The stronger interpretation of van Fraassen's new position, according to which scientific realists are accused of making a rectifiable mistake (though not of irrationality) has this shortcoming—that van Fraassen hasn't been able to identify such a mistake in a non-question-begging man- ner. Let us therefore attribute to him the weaker view that there's nothing to be said that would—or should—compel realists to mend their ways, and vice versa. On this reading, van Fraassen's current view of the realism issue is that (1) realism and anti- realism are both *irreconcilable* and *irreproachable* states of opinion, and that (2) anti- realism is true. To say that realism and antirealism are irreconcilable is to say that there are no considerations of fact or logic that can—or should—persuade propo- nents of either side to switch; to say that they're irreproachable is to say that neither side has necessarily made a mistake of fact or logic in arriving at its present state of opinion. Naturally, it follows that van Fraassen's further espousal of antirealism is beyond reproach.[1]

Van Fraassen's position brings to mind two related points of view. One could combine a belief in the irreproachability and irreconcilability of the two sides with a belief in realism. This is a position toward which I strongly incline. Alternatively, one could believe that the two sides are irreconcilable and that they're both open to reproach. This is the view that Arthur Fine calls "NOA". Van Fraassen, Fine, and I all agree that the debate is irreconcilable. However, in this chapter I show that we all draw different epistemological morals from this state of affairs. We also endorse the irreconcilability hypothesis on different grounds and with different degrees of certi- tude. Van Fraassen is the least committed to irreconcilability. Indeed, it requires a fairly bold act of interpretation to attribute this view to him. On the balance, I think that the most accurate statement of his position is that he's *prepared to accept* the thesis of irreconcilability if it should prove necessary to do so. Fine, on the other hand, thinks that the thesis of irreconcilability—or at least a large part of it—can be established by an a priori argument. He claims to have provided an algorithm for turning any realist explanation of a phenomenon into a competing antirealist explanation of the same phenomenon (my criticism of this proof is in section 3.2). My own acceptance of the irreconcilability thesis stands midway between van Fraassen's mere willingness to accept it if necessary and Fine's deductive certitude. I believe that the differences between realists and antirealists are irreconcilable on the basis of a weak induction. It seems to me that a generation of intensive argumentation has not resulted in any advantage to either side. By my reckoning, all the positive arguments for either realism or antirealism stand refuted. Contra Fine, however, I also don't think that the irreconcilability of the issue is a sure thing. It's not inconceivable to me that someone might yet devise a novel argument that gives an advantage to one side or the other. But I wouldn't hold my breath.

12.3 The Epistemological Options

Here's another potential stance on the realism issue, belonging to the same family as van Fraassen's, Fine's, and mine: realism and antirealism are both irreproachable, but their differences are reconcilable. This position is, of course, incoherent. If realism and antirealism can be reconciled, then one of them, at least, must be open to reproach. This observation leads us to wonder whether some of the three positions of section 12.2 might not prove to be incoherent as well. I don't think that a mere conceptual analysis of the notions of irreproachability and irreconcilability is going to make any trouble for them. But it's possible that some of them might be incompatible with a broad epistemological principle. For example, one might object to the postulation of irreproachable irreconcilabilities on the basis of a positivist-style principle to the effect that the irreconcilability of an issue shows it to be a "pseudo-issue", which in turn establishes that it would be an epistemic error to endorse either side. In this way, contemplation of the modes of resolution available for the realism debate naturally leads to considerations of general epistemology. The question is: what sort of epistemology allows for irreproachable irreconcilabilities?

Van Fraassen sketches an epistemological position—the "new epistemology"—which is able to accommodate irreproachable irreconcilabilities. He describes the new epistemology by contrasting it with two competing epistemologies—skepticism and orthodox Bayesianism. The skeptic is represented as subscribing to the following four propositions:

(I) There can be no independent justification to continue to believe what we already find ourselves believing.

(II) It is irrational to maintain unjustified opinion.

(III) There can be no independent justification for any ampliative extrapolation of the evidence plus previous opinion to the future.

(IV) It is irrational to ampliate without justification.

Orthodox Bayesians, according to van Fraassen, differ from skeptics in rejecting (II), for their view is that it isn't irrational to assign any prior probabilities to hypotheses that we wish (with due regard for probabilistic coherence). Having done so, however, rationality constrains us to alter our opinions only by conditionalization. Since van Fraassen regards conditionalization as a species of deductive inference, this means that the Bayesian accepts (III) and (IV), as well as (I). Van Fraassen characterizes his new epistemology as the doctrine that accepts (I) and (III), but rejects (II) and (IV). In other words, nobody can justify anything, but everything is permitted. He writes:

> Like the orthodox Bayesian, though not to the same extent, I regard it as rational and normal to rely on our previous opinion. But I do not have the belief that any rational epistemic progress, in response to the same experience, would have led to our actual opinion as its unique outcome. Relativists light happily upon this, in full agreement. But then they present it as a reason to discount our opinion so far, and to grant it no weight. For surely (they argue) it is an effective critique of present conclusions to show that by equally rational means we could have arrived at their contraries?
>   I do not think it is. (1989, 179)

The connection between the new epistemology and van Fraassen's most recent stand on the realism issue is clear: if you and I are both permitted to ampliate without justification, then we may very well arrive at irreconcilable positions without incurring any epistemic blame.

In addition to skepticism and Bayesianism, van Fraassen also discusses the "traditional epistemology, with its ampliative rules of induction and inference to the best explanation" (151). With some simplification, the traditional epistemology can also be characterized by specifying which of the skeptic's four propositions are accepted or rejected. Traditional epistemology is the view that denies principles I and III, but accepts II and IV. That is to say, traditional epistemologists insist on justification for everything and are confident that such justifications are possible. From the traditionalist perspective, we have no right to adopt realism, or antirealism, or indeed any other thesis unless we can justify our choice. But if we can justify one choice, then this justification stands as a reproach to any contrary doctrine. Thus, traditional epistemology is incompatible with the view that realism and antirealism are both irreproachable and irreconcilable. It is adherence to traditional epistemology that seems to underwrite Fine's argument for NOA. According to NOA, the disputants in the realism debate should *discard* their irreconcilable differences and restrict their circle of beliefs to matters that *are* reconcilable. Fine argues for this stance only regarding the irreconcilable differences between realists and antirealists. However, the same solution is potentially available for dealing with any irreconcilabilities. Let's call the general principle of declaring irreconcilable opinions to be out of epistemic bounds by the name of generalized NOA, or GNOA. GNOA is the indicated treatment of irreconcilabilities in the traditional epistemology: to say that two hypotheses are irreconcilable is to say that there's no way to persuade proponents of either one to change their ways. But this is to say, in turn, that neither hypothesis can be justified, in which case, according to the traditional epistemology, neither of them should be accepted. For someone who subscribes to the new epistemology of van Fraassen, however, GNOA's Draconian prohibitions against irreconcilable views are supernumerary—they resolve no outstanding difficulties.

12.4 The New Epistemology versus the Bayesian Epistemology

On van Fraassen's reckoning, "orthodox Bayesianism" lies between the traditional and the new epistemologies. Traditionalists accept principles II and IV—the principles that both maintaining present opinion and ampliation on present opinion require justification—but they reject principles I and III—the skeptical principles that there can be no justification for present opinion or ampliation thereon. The new epistemology is the mirror image of traditionalism, accepting I and III, but rejecting II and IV. Bayesianism partakes of both. Like new epistemologists, Bayesians believe that nothing can be justified (principles I and III) and that it's not irrational to maintain unjustified opinion (rejection of principle II). Bayesians agree with traditionalists, however, that ampliation requires justification (principle IV). If what we're calling the traditional epistemology is really more traditional than its competitors, then the new epistemology represents a greater departure from the status quo than Bayesianism. Why does van Fraassen not content himself with being a Bayesian? The last-quoted passage, beginning with "Like orthodox Bayesians", strongly suggests that one reason for the rejection of Bayesianism is its inability to accommodate irreproachable and irreconcilable views. In this respect, Bayesianism and traditionalism are asserted to be in the same boat. To be sure, this isn't the only problem that van Fraassen sees with Bayesian epistemology. But let us, to begin with, establish that Bayesians *don't* have a problem with irreproachable irreconcilabilities.

The prima facie problem that Bayesianism faces in dealing with irreconcilable opinions is due to the convergence theorem. To be sure, prior probabilities are free (with due respect for coherence); this is what it means for Bayesians to reject principle II. But it's well known that divergent probability functions will inevitably converge on continued conditionalization on the same incoming evidence. This seems to suggest that Bayesians can't admit that rational agents may have irreconcilable differences. It's true that the convergence theorem holds only for probabilities strictly greater than zero and less than one. It thus appears to be possible for a Bayesian analysis to equate realism with the ascription of probabilities strictly between zero and one to theories, and antirealism with the ascription of zero probabilities to all theories. Under this interpretation, realists may very well come to believe in some theories, but antirealists never will.[2] Thus, the differences between realists and antirealists are irreconcilable. The problem with this Bayesian account of irreconcilability is that there's a modified Dutch book argument showing that if we assign probabilities of zero or one to contingent hypotheses, then we will accept stupid bets that we can't win but may lose.[3] The conclusion seems to be that on the Bayesian view, the differences between realists and antirealists must be reconcilable—for either their probabilities will converge or antirealists are open to the unilateral reproach of ascribing zero probabilities to contingent hypotheses. Of course, one may deny the force of Dutch book arguments. This gambit will loom large in the discussion to come. But the status of Dutch book arguments needn't be settled at this juncture, for van Fraassen himself has indicated a manner of accommodating the existence of irreconcilable differences within a Bayesian framework in a way that evades the clutches of the Dutch bookie. Here's how it goes.

As before, scientific realism is equated with the ascription to theories of prior probabilities strictly between zero and one. Depending on their conditional probabilities and the nature of the evidence, realists will sometimes find that the posterior probabilities of some theories become high enough to warrant belief. What about antirealism? The solution, due to van Fraassen himself, is to say that antirealists ascribe *vague* prior probabilities to hypotheses having to do with unobservable events (1989, 193–194). As van Fraassen notes, a probabilist can't in every circumstance ascribe *totally* vague probabilities to theories, for "if hypothesis H implies [evidence] E, then the vagueness of H can cover at most the interval $[0, P(E)]$" (194). From the viewpoint of a theoretical agnostic, this formulation has the nice consequence that

> if E . . . becomes certain, that upper limit disappears. For the most thorough agnostic concerning H is vague on its probability from zero to the probability of its consequence, and remains so when he conditionalizes on any evidence. (194)

Thus, if we begin with an agnostic attitude about T, there is no possible evidence that could impel us to believe it. Van Fraassen doesn't discuss what happens if E becomes certainly *false*. But the consequences of this eventuality are also unproblematic. If T implies E, and if E becomes almost certainly false, then $[0, P(E)]$, the probability interval associated with T, must shrink to as close to the discrete point 0 as makes no difference. The conclusion is that one will sometimes be brought to the point of believing certain theories to be false. But this is compatible with being an antirealist, since the negation of a hypothesis that posits the existence of theoretical entities doesn't itself posit the existence of theoretical entities. I have only one amendment to make to van Fraassen's proposal. If we think it irrational to ascribe discrete probabilities of either zero or one to contingent hypotheses, then we must also think it irrational to ascribe vague probability intervals that *contain* zero or one to contingent hypotheses. Thus, I would add the requirement that the intervals ascribed to theories *not* contain zero or one to the characterization of antirealism. This change has the immediate consequence that antirealists needn't worry about the modified Dutch book argument.

In sum, both the realist's and the antirealist's epistemic stances are irreproachable from a Bayesian perspective. After all, prior probabilities are free. Yet no matter how long they conditionalize on the same new evidence, there are some aspects of their theoretical opinions that will never converge. That is to say, realism and antirealism are irreconcilable. Bayesianism thus accommodates irreproachable irreconcilabilities just as well as the new epistemology does—and it does so with fewer changes in the traditional view. This cannot come as a surprise to van Fraassen, since the Bayesian treatment of irreconcilability is his own. Indeed, van Fraassen has another reason for rejecting Bayesianism and embracing the new epistemology. But he must at least be faulted for the expository infelicity of suggesting that irreconcilability is a problem for Bayesians.

12.5 The New Epistemology versus Epistemology X

The difference between the new epistemology and Bayesianism is that the former rejects, while the latter accepts, principle IV, according to which we may not ampliate without justification. The wish to accommodate irreconcilable views doesn't give us a reason for preferring either one. But van Fraassen has another, more fundamental objection to Bayesianism. The problem is that principles III and IV together, which Bayesians accept, have an unacceptable consequence. If no ampliative rules are justified (principle III), and if all ampliation requires justification (principle IV), it follows that we have only one chance to choose our prior probabilities—"at our mother's knee", as van Fraassen says—whereupon all future belief revision has to be based on this literally childish assessment of the weight of all possible evidence on all possible hypotheses. Bayesianism provides no scope for an enhanced (or diminished) appreciation of the conditional plausibility of any hypothesis. Let's call this the problem of *rigidity*. Clearly, the new epistemology doesn't suffer from the problem of rigidity, for by denying principle IV, it allows for belief revision without justification. Thus, the new epistemology manages to accommodate irreproachable irreconcilability and to avoid the problem of rigidity at the same time.

But since the rigidity problem comes from the conjunction of principles III and IV, we can also avoid it by denying III, the principle that no ampliation is justified. For if there are justified ampliations, we may hope to move beyond the opinions obtained by conditionalization on the conditional probabilities imbibed at our mother's knee. To be sure, as long as we don't know what the justified ampliations are, we won't be able to ascertain how far we can stray from our primordial opinions. But the important point is that the general principle that there *are* justified ampliations is, by itself, compatible with any degree of distancing from our epistemic starting point. Consider now the epistemological position obtained by combining principle IV with principle I, which allows us to pick our priors without justification. Call this combination *epistemology* X. It seems that epistemology X possesses the same pair of virtues as the new epistemology—the avoidance of rigidity, and scope for irreproachable irreconcilabilities. We've seen how principle IV sets us free from the problem of rigidity. How does principle I allow for irreproachable irreconcilability? The story is only a little bit more complicated than in the case of Bayesianism. As with Bayesianism, equate realism with the ascription of probabilities strictly between zero and one to theories, and equate antirealism with the ascription of vague probabilities $(0, p(E))$ to theories, where E is the theory's empirical consequences. The availability of this assignment establishes the irreproachability of realism and antirealism in both the Bayesian epistemology and epistemology X. In the Bayesian epistemology, the irreconcilability story is already finished: Bayesians never ampliate, and no amount of conditionalization will ever bring about a reconciliation between realists and antirealists. In epistemology X, however, it has to be admitted that, as far as we know, there may be ampliative rules that can effect a reconciliation. But the important point is that the conjunction of the general principles I and IV is compatible with irreconcilability. If the justified ampliative rules are strong enough to resolve all differences, we will all have reason to rejoice. But if they don't—if there are irreconcilabilities—epistemology Xers will be able to live with them as well as new epistemologists. The occurrence of irreconcilability can be explained by epistemology X as the result of applying the same justified ampliative rules to different priors.

Perhaps this is a good place to review the epistemologies we've been juggling:

Skepticism = principles I, II, III, and IV.

Bayesianism = principles I, III, and IV.

Traditional Epistemology = principles II and IV.

New Epistemology = principles I and III.

Epistemology X = principles I and IV.

Epistemology X possesses both of the virtues claimed by the new epistemology—the avoidance of rigidity, and room for irreproachable irreconcilability. Moreover, like Bayesianism, it represents a less drastic departure from the traditional epistemology than does the new epistemology. The new epistemology is a total rebellion against the uptight demand of the traditional epistemology that there be justification for *everything*. The new epistemology takes a stand for absolute epistemic freedom—"logic only apparently constrains—a little more logic sets us free" (van Fraassen 1989, 175). The new epistemology is very 1960s. Epistemology X, with its nostalgic longing for the old certitudes, is quintessentially 1990s. Xers don't permit themselves to hope that the enveloping security of the traditional epistemology can ever be recaptured—those days are gone forever. But they continue to hope for just a little bit of justification. It remains to be seen whether such a hope can be realized in these epistemically hard times. But the requirements to accommodate irreproachable irreconcilabilities and to avoid rigidity do not yet show the hope to be in vain.

However, the prospects for epistemology X are seriously endangered by an argument that looms large in *Laws and Symmetry*. Van Fraassen claims to have a deductive proof of III, the principle that there can be no justified ampliation. If he's right, then would-be epistemology Xers will be required to add III to their system, which already contains I and IV. But this would transform them into Bayesians, whereupon they would fall prey to the problem of rigidity. If van Fraassen's proof of III is sound, epistemology X is a nonstarter. Is the proof sound? Van Fraassen shows that if we follow any ampliative rule whatever, we render ourselves liable to a dynamic Dutch book. This result might appear to establish IV, as well as III—for if all ampliative rules render us susceptible to a Dutch book, then isn't it irrational to ampliate with or without justification? And then doesn't the new epistemology collapse into Bayesianism along with epistemology X? Van Fraassen avoids this collapse by noting that we can ampliate without getting into trouble with Dutch bookies as long as we don't rely on any rules—each ampliation must be sui generis. But of course, if our ampliations are sui generis, then there's no question of providing justification for them. Thus, this avenue of escape from the clutches of Dutch bookies, while available to new epistemologists, is closed to Xers.

As noted, van Fraassen's thesis—that adherence to any ampliative rule renders us liable to a dynamic Dutch book—has the form of a deductive proof. There's no denying it. Moreover, having given the argument, van Fraassen thenceforth writes as though it had been established beyond any doubt that justified ampliation is out of the picture. But this is a bit fast. The deductive certainty of the conclusion *that adherence to any ampliative rule renders one susceptible to a dynamic Dutch book* doesn't automatically attach to the further conclusion *that it's irrational to adhere to any ampliative rule.* To arrive at the latter, we need the premise that invulnerability to dynamic Dutch books is a necessary condition for rationality. But this premise is certainly not proven by van Fraassen's argument. Van Fraassen spends no discernible effort in helping the reader to make the transition from Dutch-book vulnerability to irrationality. In fact, he regularly uses the term "incoherence", with its overwhelming connotation of irrationality, to *mean* Dutch-book vulnerability.[4] Yet the connection between the two has frequently been questioned.[5] In particular, the force of *dynamic* Dutch book arguments has been subjected to a devastating critique by Christensen (1991). Christensen notes that the mere certainty of incurring a loss, come what may, is not by itself sufficient for an indictment of irrationality. This is shown by the fact that agents whose rationality no one would be tempted to impugn may find themselves in circumstances where they are sure to lose on bets that they consider to be fair. Suppose, for example, that two people with a joint savings account disagree over the probability of some future event. A clever bookie can offer them both bets that they consider to be fair but that will result with certainty in a net diminution of their savings account. But we do not thereby conclude that it's irrational ever to disagree with one's spouse or to hold a joint savings account. Evidently, if we wish to regard vulnerability to the usual dynamic Dutch books as symptomatic of irrationality, we need to isolate the telling difference between the usual dynamic Dutch books and the foregoing "double agent Dutch book". Christensen rehearses—and refutes—all the prima facie candidates. Until a viable candidate is put forth, there is no argument against the adoption of ampliative rules. And that means, in turn, that epistemology X is open for business, at least for the time being.

After I had already written a preliminary version of this chapter, a new article by van Fraassen appeared that casts the foregoing discussion in a new light (van Fraassen 1995). In this article, van Fraassen repudiates the use of Dutch book vulnerability as a necessary condition for rationality: "I do not want to rely on Dutch Book arguments any more, nor discuss rationality in the terms they set" (9). This remark seems immediately to nullify his critique of epistemology X. But the story is a bit more complicated. Van Fraassen's new article is essentially a defense of his previously enunciated principle of "reflection".[6] This principle imposes a certain kind of temporal consistency on rational opinion. Roughly, reflection requires that our present opinion on any matter not diverge from our present opinion of what our future opinion will be. This is, as van Fraassen is acutely aware, a very counterintuitive proposal. For example, it brands as irrational the present opinion that the next time one gets drunk, one will overestimate one's capacity to drive home safely. For all its counterintuitiveness, however, van Fraassen (1984) had earlier shown that violations of reflection render us vulnerable to a dynamic Dutch book. In the new article, having renounced Dutch book arguments, he sets out to provide an alternative defense of reflection. But now that Dutch book arguments are out of the picture, his 1989 prohibition against ampliative rules is also left dangling without any visible means of support. Since the prohibition against ampliative rules is a cornerstone of the new epistemology, one would have supposed that van Fraassen would be concerned to provide an alternative defense for it as well as for reflection. But the new article has nothing to say about the 1989 claim. Strictly speaking, there is currently no argument in print that supports the prohibition of ampliative rules.

Of course, it's unlikely that the principle of reflection and the prohibition against ampliative rules will turn out to be logically independent claims. Unfortunately, van Fraassen nowhere spells out the exact nature of their logical connection. He does give an argument to show that if one never ampliates, then one automatically adheres to reflection (1995, 17–18). But ampliation here covers both adherence to ampliative rules and on-the-fly ampliations of the type sanctioned by the new epistemology. Apparently, the connection between reflection and the prohibition against ampliative rules is left as an exercise for the reader. To my great convenience, it isn't necessary to have worked through this exercise to achieve my present purpose, which is to establish that the prohibition against ampliative rules is currently without support. Now that Dutch book arguments have been withdrawn, the only extant support for the prohibition is whatever might be inherited from the new argument for reflection by virtue of the logical connection between the two doctrines. To be sure, there is some uncertainty as to what this connection may be. But this uncertainty is inconsequential, for the simple reason that there's nothing to inherit. In his 1995 article, van Fraassen does not offer any positive arguments for the adoption of reflection. He merely counters the argument that we *should not* adopt reflection because of its counterintuitiveness. He suggests that this counterintuitiveness could be due to our thinking of reflection as a semantic constraint on opinion, when in fact it may be a pragmatic constraint on assertability. He likens the violation of reflection to Moore's Paradox — "It's raining, but I don't believe it". Semantically, this statement is unobjectionable — indeed, one must confess that it may very well be true. Thus, a prohibition against statements of this type might appear to be highly counterintuitive. After all, how can we disallow an opinion that might be true? But the problem with Moore's paradox is that it commits a pragmatic infelicity: regardless of whether the statement is true, no one could ever be in a position of rationally asserting that it's true. Van Fraassen suggests that violations of reflection have the same character. Thus, the fact that opinions that violate reflection are sometimes obviously true does not settle the issue. These remarks of van Fraassen's succeed in their aim of defusing a potent prima facie objection to the reflection principle. But they do not yet constitute a strong case for its adoption. To have a positive case, van Fraassen would have to offer a derivation of reflection from pragmatic principles that command widespread assent, or at least from principles that receive independent support from their capacity to solve other problems. But the only pragmatic principle that he enunciates is the one he wants: reflection itself. This doesn't count as giving an argument for its adoption. In brief, Van Fraassen's discussion is essentially defensive and programmatic. His point is well taken: the counterintuitiveness of reflection doesn't settle the issue against it, and the program of pragmatic explanation is a viable one. But these admissions don't create any dilemmas for those who wish to promote views that are incompatible with reflection. And so epistemology X is still in business, even under the worst-case assumption that the principle of reflection entails that there can be no justified ampliative rules.

12.6 The Last Word

The time has come to reconnect our disquisition into general epistemology with the realism issue and to draw what morals there are to be drawn. I agree with van Fraassen and Fine that the differences between realists and antirealists may very well be irreconcilable. I further agree with van Fraassen that realism and antirealism may nevertheless be irreproachable. My belief in irreconcilability is based on an admittedly weak induction over the historical failures of numerous attempts at reconciliation. Similarly, my belief in their irreproachability is based on nothing more than the fact that no one has managed to formulate a non-question-begging reproach. So, one side may yet turn out to be demonstrably right in this debate. But the history of the subject suggests that it would be prudent to have an epistemology on hand that can accommodate irreproachable irreconcilabilities. Let's review how this thesis fares in the several epistemologies that we've explored.

To begin with, there can be no irreproachable irreconcilabilities in the traditional epistemology that demands universal justification. This is why Fine endorses NOA. Irreproachable irreconcilabilities are possible in Bayesianism, the new epistemology, and epistemology X, but the type of possibility is different in each case. In Bayesianism, it's deductively certain that realists and certain types of antirealists—those who ascribe vague prior probabilities to all theories—will never agree. Irreconcilability is possible in Bayesianism in the sense that these irreconcilable priors are available—somebody might select them. In fact, if the truth of Bayesianism could be established on independent grounds, we would have the missing proof of the claim that realism and (vague-probability) antirealism are definitely irreconcilable. In the new epistemology, the possibility of irreconcilability comes from the double license to choose our priors and to ampliate without justification. In this case, as well, irreconcilability is possible in the sense that two people may freely choose not to be reconciled. But there is a difference. As long as they adhere to the Bayesian epistemology, realists and antirealists can never be reconciled. In the new epistemology, however, there's nothing to stop unreconciled realists and antirealists from freely ampliating their way to their opponent's view. Thus, irreconcilability in the new epistemology is a very contingent matter—it persists as long as the disputants choose to maintain it. Finally, in epistemology X, as in Bayesianism, the possibility of irreconcilability comes only from the license to choose our priors. In this case, irreconcilability is possible in the sense that an unproved and unrefuted mathematical hypothesis is possibly true and possibly false. Since it's conceived that there may be justified ampliative rules, it is possible that one such rule will bridge the gap between realists and vague-probability antirealists. On the other hand, it's possible that there is no justified rule that will effect such a reconciliation. It all depends on what the rules are. Given our present state of near-total ignorance about the nature of these rules, it can be said that irreconcilabilities are possible in epistemology X.

Of course, there are other reasons for choosing epistemological principles besides their ability to accommodate irreproachable irreconcilabilities. In particular, the problem of rigidity seems to be adequate grounds for rejecting the narrow thesis that van Fraassen calls Bayesianism. Van Fraassen further claims that epistemology X can be eliminated on the grounds of incoherence. But we've seen that the only argument ever given for this claim has been withdrawn. So, the new epistemology and epistemology X are both still in the running. As between these two, the latter departs less drastically from traditional epistemological principles. If we take the view that conservatism in epistemic matters is a virtue, then epistemology X is, ceteris paribus, to be preferred. On the other hand, though its coherence has not successfully been challenged, it is true that epistemology X has no definite rules to offer. Xers can do nothing more than wave their hands in the direction of rules with vague slogans like "inference to the best explanation". Still, they have not yet been completely deprived of their hopes.

Notes

Chapter 1

1. Remember that the formula "X exists" is only a placeholder. Stated more fully, phenomenalism is the doctrine that only sense-data have the horizontal property—whatever that property may be—that commonsense realism ascribes to sticks and stones.

2. This argument is discussed in chapters 9–11.

3. Scientific antirealism is the doctrine (2 & –3). A disproof of antirealism is therefore a proof of –(2 & –3), or (–2 ∨ 3). If we presuppose the truth of proposition 2, the truth of 3 immediately follows. Thus, the disproof of antirealism is parlayed into a proof of realism. A similar argument establishes that the presupposition of 2 turns a disproof of realism into a proof of antirealism.

4. Here, and in the passage by Clendennin that follows, I've altered the symbolic notation of the originals for the sake of uniformity.

5. There's also the possibility of positing different degrees of reality. But this avenue has not been much traveled in recent times.

Chapter 2

1. Does this mean that a theory that has empirically equivalent rivals can't be successful? No. It means that all empirically equivalent theories must be regarded as equally successful. Since self-validation depends on social processes that are independent of the intrinsic properties of theories, the social constructionist view would have to be that *all* theories are equally successful—which comes to the same thing as saying that the concept of success doesn't apply to theories.

2. I'm uncertain whether to regard this analysis as a criticism of Laudan's confutation or as a clarification. My point is that Laudan's argument about truth doesn't make any serious difficulties for a minimal epistemic realism that claims only that it's possible to have a rationally warranted belief in theoretical truths. Laudan's argument does, however, make problems for stronger forms of epistemic realism, such as the view that we're *already* entitled to believe in some theories—and this may have been all that Laudan wished to claim. My point is that this is all that he *can* claim.

Chapter 3

1. But can't we believe things for pragmatic reasons? What about Pascal's wager? This issue is aired in chapter 8. For the present, it suffices to note the distinction between beliefs adopted on pragmatic grounds and beliefs adopted on epistemic grounds, and to stipulate that the present disquisition refers only to the latter. The claim is that there can be no pragmatic justification for epistemically based beliefs. As will be shown in chapter 8, antirealists have no stake in denying that we may have pragmatic reasons for believing in theories. It's the epistemically based belief in theories that's at issue in the debate about realism.

Chapter 4

1. On the other hand, it's conceivable that one might assign a higher probability to T_3 than to T_1 & T_2 on the grounds that T_3 is more beautiful, and then discover that T_3 is logically equivalent to T_1 & T_2. I would consider this to be a special case of grand unification.

2. See Kukla (1990b) for a fuller critique.

Chapter 5

1. Suppose T' is true, that is, T^* is true and T is false. Then, by hypothesis, there must be another theory U that has the same empirical consequences as T, that is, $U^* = T^*$. But then T' is not parasitic on T, since T' can be described as $(U^*$ & $-T)$. The remaining occurrence of "T" in "$(U^*$ & $-T)$" is trivially eliminable by replacing it with whatever specific claim about T we disagree with—for example, by claiming that one of T's theoretical entities does not exist.

2. In a personal communication, van Fraassen has claimed that $T!$ is not empirically equivalent to T, since T and $T!$ differ in their predictions of phenomena that are *observable*, even though these phenomena will never be observed. The issue is not so clear-cut. What is "observable" for van Fraassen is what it's *nomologically* possible to observe. It could be argued that if $T!$ is the true theory of the universe, then it's a consequence of scientific law that we will never observe the events governed by T_2—these events are as observationally unavailable to us as ultraviolet radiation. My analysis doesn't depend on this point, however. Even if $T!$ is judged not to be empirically equivalent to T, it would still be capable of playing the same role in the underdetermination argument as genuine vanFraassian empirical equivalents. We can be sure on a priori grounds that nothing will ever be observed that favors T over $T!$ (or vice versa); therefore, we can never come to know that T and $T!$ have different degrees of empirical support.

3. In response to an earlier version of this discussion, Leplin has maintained that my description of parasitism doesn't capture the notion that he and Laudan had in mind. According to Leplin, the fact that T' makes reference to T is incidental to its parasitism:

> Kukla wants to eliminate reference to T by directly specifying what the empirical consequences are to be. But the determination of what to specify can only be made by reference to T. That is the point of the charge of parasitism. Whether or not reference to T is made in identifying its purported rival is not the proper test of parasitism. (1997, 160)

But on this account, parasitism is plainly too strong a requirement for genuine theoryhood. Suppose scientists wish to discover a unified theory T_3 that explains the same range of empirical facts as two presently disparate theories T_1 and T_2. In this case, too, "the determination of what to specify can only be made by reference to" the conjunction T_1 & T_2. Thus, the unifying theory may be parasitic on T_1 & T_2. To be sure, most unifying theories turn out to have empirical consequences that diverge from those of the theories that they unified. But it can't be supposed that the existence of such divergences is a requirement for genuine hypothesishood.

4. Similar remarks apply to T'. Both T and T' entail that the empirical consequences of T are true. T' additionally entails that some of T's theoretical claims are false. But, obviously, T entails that *none* of its own theoretical claims are false. Once again, there's no way get from T' to T by deletion.

5. It's true that I've modified Barrow's description of A(T) in several ways, but none of these modifications would make A(T) *less* intelligible than the original version. It's also worth noting that Barrow quotes a passage by another astronomer, Frank Tipler, who also accepts A(T) as a genuine rival hypothesis. This proves that the intuitive charm of A(T) is more than a *folie à deux*.

6. I've expanded on this theme elsewhere (Kukla, 1995).

7. Note also that McMichael's treatment requires us to give theories a syntactic, as opposed to semantic, characterization.

8. The proof of (1) and (2) from (3) requires the assumption that nothing is both an emerald and a sapphire. But this is an item of information that we want in our account of the world anyway—and it must either be added to the conjunction of (1) and (2), or be derived from the equally potent assumption that nothing is both green and blue.

9. Van Fraassen's position on this issue is that the empirical content of T *is* finitely axiomatizable—in English. The empirical content of T is fully captured by a single axiom: "The empirical consequences of T are true" (1980, 46–47). There are problems with this claim, however, which will loom large in chapter 10.

Chapter 6

1. Page references in this section are to the reprinted version of Laudan and Leplin's article in Laudan (1996)

Chapter 7

1. Recall that Leplin and Laudan (1993) also appealed to parasitism as a means of eliminating certain hypotheses (see chapter 5). Clendinnen's inferences from parasitism are different from Leplin and Laudan's, however. Leplin and Laudan cite the parasitism of T* as a reason for denying that it's a proper theory; Clendinnen cites the parasitism of T^ as a reason for supposing that belief in T^ commits us to belief in T.

2. I've altered Clendinnen's notation to conform to mine. Clendinnen defines Te as "the theory of electrons" and Te' as "the hypothesis that there are no electrons but observable phenomena are just as they would be if there were" (1989, 86). In the passage quoted, I replace Te with T, Te' with (T^ & not-T), and (Te or Te') with T^. A better translation of Clendinnen's words would probably be obtained with the substitution of T* or T# for T^. But, as noted earlier the argument can only be strengthened by running it on T^.

Chapter 9

1. It could be argued that "all" is stronger than necessary here—that the impossibility of theory-neutral observation already follows from the narrower thesis that the cognitive differences that are implicated in the choice among competing theories have an effect on perception. This hypothesis is still far broader than any that the New Look experiments can reasonably be said to have addressed. Moreover, the supposedly narrower thesis may be equivalent to the stronger one in the end, for it's plausible that for *any* cognitive difference, there is some theoretical choice in which this difference is implicated.

2. Is the output of P propositional? Does P provide C with the proposition that it looks as though there's an apple on the desk, or does it pass along a picture of an apple on a desk, which C then converts into a proposition on its own? This is the kind of detailed scientific question that can only be settled by a program of empirical research. One thing is certain, however: for perception to influence belief fixation, its content has to be converted to a proposition *somewhere* along the line. Moreover, if it's to provide us with a pool of commensurable beliefs, at least part of the conversion to propositions has to be performed by mechanisms that function in the same manner in all of us. For present purposes, it will do no harm to allocate this function to P itself.

3. It might be maintained that this "all" is too strong—that the necessary condition for incommensurability is only that our theories influence those of our perceptions that confirm or disconfirm our theories. I suspect that this weakened claim is equivalent to the original, for it seems overwhelmingly likely that every perception is epistemically relevant to some possible theoretical choice. Whether or not this is so, the thesis that Churchland argues for does not, by itself, produce an incommensurability problem. To produce an incommensurability problem, you have to establish at least that every theory influences *all the perceptions that are relevant to its own confirmational status*. But the hypothesis that Churchland tries to establish is only that every theory influences some perceptions. Compare note 1, which makes essentially the same point about the previous universal quantifier.

4. Granny distinguishes here between two processes of belief fixation rather than two categories of scientific hypotheses. These distinctions don't line up in a one-to-one fashion, for we may learn the truth of an observation statement by a process of inference, as when we induce that the emeralds we haven't seen are green. But it's clear what Granny's distinction between hypotheses would be: that a hypothesis is observational just in case it *can* be established noninferentially.

Chapter 11

1. Of course, as noted earlier, there's always the possibility that a *fourth* distinction might be devised that vindicates antirealism. Thus, it might be considered an overstatement to claim that the rejection of hypothesis (2) would eliminate antirealism from consideration. But the rejection of hypothesis (2) would constitute as strong a case against antirealism as there can presently be.

Chapter 12

1. On the other hand, there may well be an incompatibility between van Fraassen's new position and the vulnerability criterion of belief that he espoused in 1985. In 1985, van Fraassen expressed his disdain for realists who allow themselves the luxury of believing in hypotheses that will never have to undergo the test of future experience. In 1989 he seems willing at once to admit that the debate between realists and antirealists is irreconcilable and to espouse antirealism. But if the realism issue is irreconcilable, then isn't the avowal of either side a particularly flagrant case of "empty strutting"? Shouldn't his aversion to empty strutting lead him to adopt the NOA?

2. I equate belief in X with the ascription of a very high probability for X.

3. For more detail, see Salmon (1988).

4. See, for example, p. 169 of *Laws and Symmetry*.

5. See Eells (1982) for references to a number of critiques of Dutch book arguments.

6. This principle was introduced in an earlier article (van Fraassen 1984). In the new essay, van Fraassen is actually concerned with establishing a generalized version of the earlier principle. But the differences between the two versions don't figure in my discussion.

Bibliography

Barrow, J. D. (1991). *Theories of everything*. New York: Oxford University Press.

Benacerraf, P. (1965). What numbers could not be. *Philosophical Review* 74, 47–73.

———. (1973). Mathematical truth. *Journal of Philosophy* 70, 661–679.

Boyd, R. N. (1984). The current status of scientific realism. In J. Leplin (ed.), *Scientific realism*. Berkeley: University of California Press (pp. 41–82).

Bruner, J. (1957). On perceptual readiness. *Psychological Review* 64, 123–152.

Cartwright, N. (1983). *How the laws of physics lie*. Oxford: Clarendon Press.

Chomsky, N. (1980). Recent contributions to the theory of innate ideas: Summary of oral presentation. In H. Morick (ed.), *Challenges to empiricism*. Indianapolis, Ind.: Hackett (pp. 230–239).

Christensen, D. (1991). Clever bookies and coherent beliefs. *Philosophical Review* 100, 229–247.

Churchland, P. M. (1985). The ontological status of observables: In praise of superempirical virtues. In P. M. Churchland & C. A. Hooker (eds.), *Images of science*. Chicago: University of Chicago Press (pp. 35–47).

———. (1988). Perceptual plasticity and theoretical neutrality: A reply to Jerry Fodor. *Philosophy of Science* 55, 167–187.

Clendinnen, F. J. (1989). Realism and the underdetermination of theory. *Synthese* 81, 63–90.

Creath, R. (1985). Taking theories seriously. *Synthese* 62, 317–345.

Cushing, J. T. (1991). Quantum theory and explanatory discourse: Endgame for understanding? *Philosophy of Science* 58, 337–358.

Dennett, D. C. (1971). Intentional systems. *Journal of Philosophy* 68, 87–106.

Dirac, P. A. M. (1963). The evolution of the physicist's picture of nature. *Scientific American* 208, 45–53.

Eells, E. (1982). *Rational decision and causality*. Cambridge: Cambridge University Press.

Ellis, B. (1985). What science aims to do. In P. M. Churchland & C. A. Hooker (eds.), *Images of science*. Chicago: University of Chicago Press (pp. 48–74).

Feyerabend, P. K. (1969). Science without experience. *Journal of Philosophy* 66, 791–794.

Field, H. (1980). *Science without numbers*. Oxford: Blackwell.

Fine, A. (1984). The natural ontological attitude. In J. Leplin (ed.), *Scientific realism*. Berkeley: University of California Press (pp. 83–107).

——. (1986). Unnatural attitudes: Realist and instrumentalist attachments to science. *Mind* 95, 149–179.

Fodor, J. (1981). The present status of the innateness controversy. In J. Fodor, *Representations: Philosophical essays on the foundations of cognitive science*. Cambridge: MIT Press (pp. 257–316).

——. (1983). *The modularity of mind*. Cambridge: MIT Press.

——. (1984). Observation reconsidered. *Philosophy of Science* 51, 23–43.

——. (1988). A reply to Churchland's "Perceptual plasticity and theoretical neutrality." *Philosophy of Science* 55, 188–198.

Foss, J. (1984). On accepting van Fraassen's image of science. *Philosophy of Science* 51, 79–92.

Friedman, M. (1974). Explanation and scientific understanding. *Journal of Philosophy* 71, 5–19.

——. (1982). Review of van Fraassen (1980). *Journal of Philosophy* 79, 274–283.

——. (1983). *Foundations of space-time theories*. Princeton, N.J.: Princeton University Press.

Giere, R. N. (1985a). Constructive realism. In P. M. Churchland & C. A. Hooker (eds.), *Images of science*. Chicago: University of Chicago Press (pp. 75–89).

——. (1985b). Philosophy of science naturalized. *Philosophy of Science* 52, 331–356.

Gilman, D. (1992). What's a theory to do . . . with seeing? Or some empirical considerations for observation and theory. *British Journal for the Philosophy of Science* 43, 287–309.

Glymour, C. N. (1980). *Theory and evidence*. Princeton, N.J.: Princeton University Press.

——. (1984). Explanation and realism. In J. Leplin (ed.), *Scientific realism*. Berkeley: University of California Press (pp. 173–192).

Good, I. J. (1968). Corroboration, explanation, evolving probability, simplicity and a sharpened razor. *British Journal for the Philosophy of Science* 19, 123–143.

Goodman, L. E. (1992). The trouble with phenomenalism. *American Philosophical Quarterly* 29, 237–252.

Hacking, I. (1983). *Representing and intervening*. Cambridge: Cambridge University Press.

Hardin, C. L., & Rosenberg, A. (1982). In defense of convergent realism. *Philosophy of Science* 49, 604–615.

Harman, G. (1986). *Change in view: Principles of Reasoning*. Cambridge: MIT Press.

Hausman, D. M. (1982). Constructive empiricism contested. *Pacific Philosophical Quarterly* 63, 21–28.

Hoefer, C., & Rosenberg, A. (1994). Empirical equivalence, underdetermination, and systems of the world. *Philosophy of Science* 61, 592–607.

Horwich, P. G. (1982). Three forms of realism. *Synthese* 51, 181–201.

——. (1991). On the nature and norms of theoretical commitment. *Philosophy of Science* 58, 1–14.

Kuhn, T. S. (1962). *The structure of scientific revolutions*. Chicago: University of Chicago Press.

Kukla, A. (1990a). Evolving probability. *Philosophical Studies* 59, 213–224.

——. (1990b). Ten types of theoretical progress. In A. Fine, M. Forbes, & L. Wessels (eds.), *PSA 1990*. East Lansing, Mich.: Philosophy of Science Association (pp. 457–466).

——. (1993). Laudan, Leplin, empirical equivalence, and underdetermination. *Analysis* 53, 1–7.

——. (1994). Some limits to empirical inquiry. *Analysis* 54, 153–159.

——. (1995). Is there a logic of incoherence? *International Studies in the Philosophy of Science* 9, 57–69.

Langley, P., Simon, H., Bradshaw, G., & Zytkow, J. (1987). *Scientific Discovery*. Cambridge: MIT Press.

Latour, B., & Woolgar, S. (1979). *Laboratory life: The social construction of scientific facts*. London: Sage.

Laudan, L. (1977). *Progress and its problems*. Berkeley: University of California Press.

———. (1981). A confutation of convergent realism. *Philosophy of Science 48*, 19–49.

———. (1984a). Explaining the success of science: Beyond epistemic realism and relativism. In J. T. Cushing, C. F. Delaney, & G. M. Gutting (eds.), *Science and reality: Recent work in the philosophy of science*. Notre Dame, Ind.: University of Notre Dame Press (pp. 83–105).

———. (1984b). *Science and values*. Berkeley: University of California Press.

———. (1996). *Beyond positivism and relativism: Theory, method and evidence*. Boulder, Col.: Westview Press.

Laudan, L., & Leplin, J. (1991). Empirical equivalence and underdetermination. *Journal of Philosophy 88*, 449–472.

Leeds, S. (1994). Constructive empiricism. *Synthese 101*, 187–221.

Leplin, J. (1987). Surrealism. *Mind 96*, 519–524.

———. (1997). *A novel defense of scientific realism*. New York: Oxford University Press.

Leplin, J., & Laudan, L. (1993). Determination underdeterred. *Analysis 53*, 8–15.

Maxwell, G. (1962). The ontological status of theoretical entities. In H. Feigl & G. Maxwell (eds.), *Scientific explanation, space and time*. Minneapolis: University of Minnesota Press (pp. 3–27).

McAllister, J. W. (1993). Scientific realism and criteria for theory-choice. *Erkenntnis 38*, 203–222.

McMichael, A. (1985). Van Fraassen's instrumentalism. *British Journal for the Philosophy of Science 36*, 257–272.

Morrison, M. (1990). Unification, realism and inference. *British Journal for the Philosophy of Science 41*, 305–332.

Musgrave, A. (1985). Realism versus constructive empiricism. In P. M. Churchland & C. A. Hooker (eds.), *Images of science*. Chicago: University of Chicago Press (pp. 197–221).

Psillos, S. (1996). On van Fraassen's critique of abductive reasoning: Some pitfalls of selective skepticism. *The Philosophical Quarterly 46*, 31–47.

Putnam, H. (1975a). *Mathematics, matter, and method: Philosophical papers* (vol. 1). Cambridge: Cambridge University Press.

———. (1975b). *Mind, language and reality: Philosophical papers* (vol. 2). Cambridge: Cambridge University Press.

———. (1978). *Meaning and the moral sciences*. London: Routledge & Keegan Paul.

———. (1980). The "innateness hypothesis" and explanatory models in linguistics. In H. Morick (ed.), *Challenges to empiricism*. Indianapolis: Hackett (pp. 240–250).

Quine, W. V. O. (1975). On empirically equivalent systems of the world. *Erkenntnis 9*, 313–328.

Rosenberg, A. (1983). Protagoras among the physicists. *Dialogue 22*, 311–317.

Salmon, W. C. (1988). Dynamic rationality: Propensity, probability, and credence In J. H. Fetzer (ed.), *Probability and causality: Essays in honor of Wesley C. Salmon*. Dordrecht: Reidel (pp. 3–40).

———. (1989). Four decades of scientific explanation. In P. Kitcher & W. C. Salmon (eds.), *Minnesota studies in the philosophy of science* (vol. 13). Minneapolis: University of Minnesota Press (pp. 3–219).

Sober, E. (1990). Contrastive empiricism. In C. W. Savage (ed.), *Minnesota studies in the philosophy of science* (vol. 14). Minneapolis: University of Minnesota Press (pp. 392–410).

Trout, J. D. (1992). Theory-conjunction and mercenary reliance. *Philosophy of Science 59*, 231–245.

van Fraassen, B. C. (1980). *The scientific image.* Oxford: Clarendon Press.

——. (1983). Glymour on evidence and explanation. In J. Earman (ed.), *Minnesota studies in the philosophy of science* (vol. 10). Minneapolis: University of Minnesota Press (pp. 165–176).

——. (1984). Belief and the will. *Journal of Philosophy 81,* 235–256.

——. (1985). Empiricism in the philosophy of science. In P. M. Churchland & C. A. Hooker (eds.), *Images of science.* Chicago: University of Chicago Press (pp. 245–308).

——. (1989). *Laws and symmetry.* Oxford: Clarendon Press.

——. (1995). Belief and the problem of Ulysses and the sirens. *Philosophical Studies 77,* 7–37.

Worrall, J. (1984). An unreal image. *British Journal for the Philosophy of Science 35,* 65–80.

Index